T0211117

# SpringerBriefs in Applied Sciences and Technology

## Multiphase Flow

*Series editors*

Lixin Cheng, Portsmouth, UK
Dieter Mewes, Hannover, Germany

For further volumes:
http://www.springer.com/series/11897

Victor I. Terekhov · Maksim A. Pakhomov

# Flow and Heat and Mass Transfer in Laminar and Turbulent Mist Gas-Droplets Stream over a Flat Plate

 Springer

Victor I. Terekhov
Maksim A. Pakhomov
Laboratory of Thermal and Gas Dynamics
Kutateladze Institute of Thermophysics,
  Siberian Branch of Russian Academy
  of Sciences
Novosibirsk
Russia

ISSN  2191-530X　　　　　ISSN 2191-5318　(electronic)
ISBN  978-3-319-04452-1　　ISBN 978-3-319-04453-8　(eBook)
DOI 10.1007/978-3-319-04453-8
Springer Cham Heidelberg New York Dordrecht London

Library of Congress Control Number: 2014932287

Printed on acid-free paper

Springer is part of Springer Science+Business Media (www.springer.com)

# Acknowledgments

This work was partially supported by the President's Fund for Young Scientists (Grant MD-670.2012.8) and by the Russian Foundation for Basic Research (Project 11-08-00112a).

# Contents

# Nomenclature

| | |
|---|---|
| $C_D$ | Drag coefficient |
| $C_p$, $C_{pA}$, $C_{pL}$, $C_{pV}$ | Heat capacities of gas–vapor mixture, air, liquid, and vapor, correspondingly, J kg$^{-1}$ K$^{-1}$ |
| $d$ | Droplet diameter, m |
| $D$ | Vapor diffusion coefficient in air, m$^2$ s$^{-1}$ |
| $D_{xL}$, $D_{rL}$ | Turbulent diffusivities of droplets in the longitudinal and transverse directions due to the stochastic motion of droplets and its entrainment into the gas flow by intense vortices, m$^2$ s$^{-1}$ |
| $J$ | Steam mass flux from the surface of evaporating droplet, J kg$^{-1}$ m$^{-2}$ |
| $k$ | Turbulent kinetic energy, m$^2$ s$^{-2}$ |
| $K_A$, $K_V$ | Mass fraction of air and steam in the case of binary air–vapor mixture |
| $L$ | Heat of evaporation, J kg$^{-1}$ |
| $\mathrm{Lp} = \rho d \sigma_L / \mu^2$ | Laplace number |
| $M_L$, $M_V$ | Mass fraction of droplets and steam in the case of triple air–liquid–vapor mixture |
| Nu | Nusselt number |
| Nu$_A$ | Nusselt number in the single-phase flow |
| $P$ | Pressure, Pa |
| $\bar{R}$ | Gas constant, J kg$^{-1}$ K$^{-1}$ |
| $T$ | Temperature, K |
| $\langle t u_j \rangle$ | Turbulent heat flux, K m s$^{-1}$ |
| $U_j$ | Velocity components in longitudinal and transverse directions, m s$^{-1}$ |
| $U_1$ | Gas velocity value in the free-stream flow, m s$^{-1}$ |
| $V$ | Transversal velocity component, m s$^{-1}$ |
| $X$ | Flat plate length, m |
| $U_+ = U/U_*$ | Dimensionless velocity |
| $U_*$ | The friction velocity, m s$^{-1}$ |
| $\langle u^2 \rangle$, $\langle v^2 \rangle$ | r.m.s. velocity fluctuations of the gas phase in axial and radial directions, m$^2$ s$^{-2}$ |

| | |
|---|---|
| $\langle u_i u_j \rangle$ | Reynolds stresses, m$^2$ s$^{-2}$ |
| $x_j$ | Coordinates, m |
| $x$ | Longitudinal coordinate, m |
| $y$ | Transversal coordinates, m |
| $y_+ = y U_* / v$ | Dimensionless distance from the wall |
| Pr | Prandtl number, $C_P \mu / \lambda$ |
| Re | Reynolds number, $U_\infty x / v$ |
| Re$_L$ | Reynolds number of the dispersed phase, |

$$d\sqrt{(U - U_L)^2 + (V - V_L)^2} \Big/ v$$

| | |
|---|---|
| Sc | Schmidt number, $v/D$ |
| St | Stokes number, $\tau/\tau_f$ |
| Tu | Turbulence level |
| We $= |\vec{U} - \vec{U}_L|^2 \rho d / \sigma_L$ | Weber number |

## Greek Symbols

| | |
|---|---|
| $\Phi$ | Volume concentration of droplets |
| $\alpha$ | Heat transfer coefficient, W m$^{-2}$ K$^{-1}$ |
| $\delta*$ | Displacement thickness, $\delta^* = \int_0^\delta \left(1 - \frac{\rho U}{\rho_1 U_1}\right) dy$, m |
| $\delta**$ | Momentum thickness, $\delta^{**} = \int_0^\delta \frac{\rho U}{\rho_1 U_1}\left(1 - \frac{U}{U_1}\right) dy$, m |
| $\tilde{\varepsilon}$ | Dissipation of the turbulent kinetic energy, m$^2$ s$^{-3}$ |
| $\mu$ | Dynamic viscosity, N s m$^{-2}$ |
| $v$ | Kinematic viscosity, m$^2$ s$^{-1}$ |
| $\langle \theta_L^2 \rangle$ | Temperature fluctuations in the dispersed phase, K$^2$ |
| $\langle \theta_L u_{Lj} \rangle$ | Turbulent heat flux in the dispersed phase, K m s$^{-1}$ |
| $\rho$ | Density, kg m$^{-3}$ |
| $\tau$ | Particle relaxation time, $\rho_L d^2/(18 \mu W)$, s |
| $\tau_f$ | Relaxation time, $X/U_1$, s |
| $\tau_\Theta$ | Particle thermal relaxation time, $C_{pL} \rho_L d^2/(12 \lambda Y)$, s |

## Subscripts

| | |
|---|---|
| 1 | Parameter under free-stream conditions |
| $A$ | Air |
| $L$ | Dispersed phase |
| $P$ | Particles |
| $T$ | Turbulent parameter |
| $V$ | Vapor |
| $W$ | Parameter under wall conditions |

# Symbol

< >   Ensemble average

# Acronym

DNS      Direct numerical simulation
LEVM     Linear eddy viscosity model
LDA      Laser Doppler anemometry
LRN      Low Reynolds number
PDF      Probability density function
TKE      Turbulent kinetic energy

# Abstract

The flow dynamics and heat and mass transfer in the laminar and turbulent boundary layer on the flat vertical plate have been numerically investigated. Wide range of factors that have an effect on heat and mass transfer and flow patterns of thermal and mass concentration profiles in the laminar and turbulent boundary layer is analyzed. In the predictions several parameters are varied: Reynolds number $Re_x = U_1 x/v = 10^3–10^6$; wall temperature $T_w = 323–473$ K; mass fraction of the water droplets $M_{L1} = 0–0.05$ and its diameter in the free-stream $d_1 = 0–100$ μm, inlet temperature of the gas-droplets mixture $T_1 = 293$ K and the mass vapor fraction in the free-stream $M_{v1} = 0.014$. The presence of the liquid phase greatly affects the temperature distribution, which ultimately causes a more rapid growth of heat fluxes, as compared to the friction. With growth in the particle mass concentration, the friction on the wall increases only slightly. The increase in droplets mass fraction results in significant intensification of the heat transfer rate compared (up to 10 times in laminar flow regime and up to 5 times in turbulent one) with the one-phase airflow. The predictions are tested against the experimental results for mist gas-droplets laminar and turbulent gas-dispersed boundary layer. Good agreement is obtained over the whole range of initial conditions.

# Chapter 1
# Flow Dynamics, Heat and Mass Transfer in Two-Phase Laminar and Turbulent Boundary Layer on a Flat Plate with and without Heat Transfer Between Solid Wall and Flow: The State-of-the-Art

In the section the results of experimental and numerical studies of disperse particles behavior and their influence on the parameters of gas flow in the laminar and turbulent boundary layers are given and analyzed. Extensive experimental and theoretical information on the dynamics and heat and mass transfer in the two-phase laminar and turbulent systems is given in the reviews (Brennen 2006; Clift et al. 1978; Crowe et al. 1998; Drew 1983; Friedlander 1977; Ishii 1975; Loth 2006; Michaelides 2006; Nigmatulin 1991; Osiptsov 1997; Soo 1967; Varaksin 2007; Zaichik et al. 2008). Note that from the analysis of the now available experimental and numerical results it is obvious that influence of the particles on the near-wall flow may be of two kinds. First, the dispersed phase may effects the flow in the boundary layer for the account of its modification. Second, the particles directly influence the flow in the boundary layer due to their inertial nature, namely for the account of the present dynamic and heat interface slip.

## 1.1 Dusty Flow in a Laminar and Turbulent Boundary Layer

Detailed measurements of the heterogeneous flow in the boundary layer were carried out in (Varaksin 2007; Varaksin et al. 1995). The boundary layer developed along the side surface of the rode with the hemispherical end. The rode was streamlined by the upward air flow. Electrocorundum $Al_2O_3$ solid particles $d = 50$ μm of were used as a dispersed phase. Mass concentration of the particles varied in the range $M_P = 0$–$0.26$. Experiments were performed with the use of LDA method. Measurements of the phase velocities and their pulsations were carried out in all areas of the boundary layer: the laminar, transitional and turbulent ones. The particles significantly affect the profile of gas velocity in the laminar boundary layer. It becomes filled up as a result of air acceleration by the particles in the wall area. The difference between phase velocities becomes maximal near the wall for the account of the particles inertia. Increase of the gas

V. I. Terekhov and M. A. Pakhomov, *Flow and Heat and Mass Transfer in Laminar and Turbulent Mist Gas-Droplets Stream over a Flat Plate*, SpringerBriefs on Multiphase Flow, DOI: 10.1007/978-3-319-04453-8_1, © The Author(s) 2014

velocity profile thickness results in the growth of its gradient on the wall that causes the growth of the wall surface friction in the laminar boundary layer. For the two-phase boundary layer it was obtained that the particles cause the earlier start of the laminar-turbulent transition. The characteristics of the gas-dispersed turbulent boundary layer were measured in the area of the flow relaxation, i.e. in the area where the particles velocity decreases in the upstream direction. The particles velocity near the wall remains much higher than the velocity of the one-phase flow. Therefore the qualitative character of gas velocities and particles distribution in the turbulent boundary layer remains similar to the one in the laminar boundary layer.

One of the first numerical works on solid particles dynamics in the laminar boundary layer on the flat plate were the chapters (Osiptsov 1980; Stulov 1979). The numerical investigation was performed with the use of the two-fluid model (Drew 1983; Ishii 1975). It was assumed that the particles are spheres with hard boundaries and their volume fraction $\Phi$ is small. The added mass, aerodynamic drag and pressure gradient were taken into account (Michaelides 2006), whereas the work (Drew 1983) considered only the force of aerodynamics drag. The initial velocities of gas and particles were taken equal. Mass fraction of the particles varied in the wide range ($M_P = 0$–10). The chapters have studied the evolution of the profiles of the longitudinal and transverse components of the mean air and particles velocities as well as the concentration of the particles in the laminar boundary layer on the semi-infinite plate and the coefficient of the surface friction. It is shown that the mean longitudinal velocity of the particles is higher than the gas velocity over the whole cross-section of the boundary layer. The gas velocity profiles in the two-phase flow are filled up compared with the one-phase flow. This occurs due to intense impulse exchange at the presence of the gradient in the phase velocities. The flow structure in the wide range change of particles mass fraction is of a qualitatively similar view. The profiles of both phases at the large distances from the beginning of the plate become self-similarity. At the small distances the transverse velocity of gas is higher than the respective value for the particles. There are areas near the wall where the particles concentration increases to infinity as approaching to the wall; unfortunately it has not been proved experimentally yet. Friction coefficient in the two-phase flow becomes much higher than in the one-phase flow.

Among the studies concern the problems of body resistance in heterogeneous flows pay attention to the experimental (Drew 1983; Varaksin 2007) and numerical (Asmolov 1993, 1995) works. The maximal effect of the particles on friction on the wall is reached at $20°$ angle of the wedge over the entire range of particles dimensions ($d = 0$–90 μm) and $M_P = 0$–0.3 (Balanin and Lashkov 1982). Critical size of particles $d \approx 30$ μm after which the further increase of the dispersed phase diameter has no effect on the friction for all angles of the wedge is shown.

In the paper (Agranat 1986) the analytical formula for the coefficient of heat and mass transfer in the non-equilibrium dusty laminar boundary layer has been obtained.

$$\mathrm{Nu}_x = \frac{C_f}{2} \mathrm{Re}_x^{1/2} \mathrm{Pr}_{eff}^{1/3} \tag{1.1}$$

In Eq. (1.1) $\mathrm{Nu}_x = \frac{x}{|T_W - T_\infty|} \frac{\partial T}{\partial y}$, $\mathrm{Pr}_{eff} = \frac{\mathrm{Pr}(1 + aC_{pP}/C_p)}{(1+\beta)}$, where $a = m_P n_{P\infty}/\rho$ is the ratio of mean densities of the gaseous and particulate phases.

In the all previous works forces of the added mass, aerodynamic drag and pressure gradient were taken into account as forces of interfacial interaction. The attempts to consider effect of Saffman and Magnus forces were made in the works (Asmolov 1992; Naumov 1990; Wang and Levy 2006). In these works it was shown that at the modified Reynolds number of the particles $\mathrm{Re}_P = U_0 d/v < 10$. The Magnus force projection on the transverse axis is much lesser that Saffman force. In the work it is shown that near the wall the transverse velocity of the particles is directed toward the wall. The longitudinal velocity of the particles in this area is higher than gas velocity; therefore, Saffman force is directed toward the plate. With the increase of the distance from the beginning of the plate the difference of the longitudinal phase velocities decreases, that results in the decrease of Saffman force value. At the absence of the interfacial slip the profile of the particles mass concentration coincides with the one obtained in (Osiptsov 1980) without taking into account the Saffman force. Increase of Reynolds number of the particle results in the growth of the transverse velocity of the particles that leads to the increase of the mass flow rate of the depositing particles. This causes the decrease of their mass concentration value near the wall and disappearance of the area with maximum mass concentration on the wall. Consideration of Saffman force leads to the decrease of drag coefficient.

The flow patterns, gas phase turbulence and particles dispersion in a turbulent gas-dispersed boundary layer is examined in (Wang and Levy 2006; Kaftori et al. 1995; Rogers and Eaton 1990, 1991; Taniere et al. 1997).

Characteristics of the turbulent gas-dispersed boundary layer are experimentally studied in the chapters (Rogers and Eaton 1990, 1991). Measurements were carried out with the use of LDA method in the upward flow on the flat vertical plate. Difference between the phase velocities is close to the value of free-fall particles velocity and is practically constant across the boundary layer. Measurement results have shown that at $M_P = 0.02$ the glass particles with $d = 50$ and $90\ \mu m$ practically have no effect on the distribution of the averaged and fluctuating gas velocities. At that the value of the longitudinal pulsations of the particles velocity with the size $d = 50\ \mu m$ is close to the respective value for gas. Pulsations of the particles velocity for $d = 90\ \mu m$ are higher than the pulsations of the gas phase. The increase of the longitudinal pulsations of the particles velocity near the wall, where the relative inertia of the particles increases for the account of the decrease of the energy-carrying time scale of turbulent eddies, is explained by non-homogenous distributions of the mean particles velocities. The transverse velocity of the particles fluctuations are lesser than the one for the gas phase over the whole boundary layer cross-section. The difference between them increases in the vicinity of the wall owing to the fact that the particles are worse carried away by the turbulent vortexes. Secondly, the transverse averaged velocity of the particles is close to null over the section of the boundary layer. Therefore, there is no reason for additional generation of the particles turbulence from the mean motion as in the case with the longitudinal pulsations.

The copper particles with the diameter of 70 μm with the mass concentration of $M_P = 0.2$ has practically no effect on the averaged characteristics of the gas flow (Rogers and Eaton 1991). Nevertheless presence of the dispersed phase effects the spectrum of the longitudinal pulsations of gas suppressing the low-frequency components. The measurements showed a strong correlation between the degree of damping and the particle concentration in the log region of the boundary layer. It was observed that the particles damped gas turbulence.

In Taniere et al. (1997), the gas and particle velocity fluctuations and the particle concentration profiles in a horizontal two-phase boundary layer on a flat plate were measured. It is shown that the intensity of longitudinal velocity fluctuations of the particles is greater than the corresponding value for the carrier phase in the entire cross section of the boundary layer. The transverse velocity fluctuations of the particles are smaller than those for the gas almost across the entire boundary layer, except for a thin near-wall region ($y_+ = yu_*/v < 50$), where $u_*$ is the dynamic velocity and $v$ is the kinematic viscosity. The maximum intensity of the longitudinal fluctuations is located in the wall region.

Particles behavior in the turbulent boundary layer of a dilute gas-dispersed flow past a 1.5 m long flat plate was studied experimentally and numerically in the chapter (Wang and Levy 2006). For the measurements was used miniature fiber optic probe and for the computations were realized with the using of CFD commercial code CFX 4. The experimental and numerical results show a non-uniform distribution of particle mass fraction with the crest of the concentration inside the boundary layer. The particles mass fraction inside the boundary layer increased and the location of the peak value occurred closer to the flat plate surface. This phenomenon is believed to be related to the Saffman force and particle–wall interaction. Particle size has significant effect on the behavior of the dispersed phase. In the boundary layer, particles velocities of the 200 μm particles is higher than those of the 60 μm particles. Mass concentration of larger particles is also greater with the peak value in the wall zone. Both measurements and computations showed that particles play important role in the mechanism of understanding of the motion of the dispersed phase in the boundary layer. Mass fraction and diameter of the particulate phase affected on particles velocity profiles can be explained by particles inertia and momentum transfer from particles to air inside the boundary layer.

Experimental and numerical investigations of the laminar downstream two-phase flow in the vertical flat plate boundary layer were provided in the (Hussainov et al. 1995; Kartushinsky et al. 2009). Gas and particles phases velocities and particles mass concentration were measured with LDA and PDA devices. It is shown that the mass fraction have maximum crest located inside the boundary layer, whereas near the wall there is no drastic increase of the particles concentration specific for the chapters (Soo 1967; Stulov 1979). The formula determining the dispersed phase deposition value onto the flat plate surface has been proposed.

The takes into account dynamics of the flow, adhesive properties of the particles and material of the plate and the probabilistic character of the particles capturing by the wall surface.

Mechanism in the transport of inertia-dominated particles was studied in (Shin et al. 2003) in the problem of particle deposition inside a turbulent boundary layer. Due to the finite inertia of particles and mean shearing of the carrier flows, the transport of inertia-dominated particles inside a turbulent boundary layer is seriously affected by a non-equilibrium memory effect. A non-equilibrium equation for the particle Reynolds stress was derived from the stochastic differential equation of motion of particles governed by the Stokes drag and shear-induced lift forces. This new model was then applied to the problem of particle deposition in the fully developed turbulent channel flows. The predicted deposition velocities as a function of particle relaxation time were in excellent agreement with existing numerical and experimental data.

Numerical study the effect of gas injection and suction on the flat plate surface was presented in the (Grishin and Zabarin 1987, 1988). The gas injection results in appearing a dust-free region in the wall zone and crossing particle tracks. The gas suction leads to particle deposition and disappearance of particle accumulation in the wall region. The similar results were obtained also in the predictions of (Chamka and Peddieson 1992).

The process of formation of liquid moving film from the melted particles depositing on the body surface is numerically considered in the work (Osiptsov and Shapiro 1989, 1993). The blunted body streamlined by the two-phase flow has been investigated. The problem of the body streamlined, the film formation and its dynamics have been couple solved. The isothermal film flowing on the sphere surface has been investigated. The film thickness, friction coefficient and heat transfer near the symmetry axis have been found for non-isothermal flow. The conditions at which the presence of the film decreases the heat flows to the wall have been determined.

In the paper (Naumov 1990) have studied the effect of the Saffman force on the distributions of the heat transfer on the flat plate. It was obtained that decrease of the maximum of $C_f \mathrm{Re}_x^{1/2}$ and $\mathrm{Nu}/\mathrm{Re}_x^{1/2}$ and, also, a displacement of those quantities over the plate length.

# References

V.M. Agranat, Reynolds analogy in a dusty laminar boundary layer. Fluid Dyn **21**, 983–985 (1986)

A.M. Grishin, V.I. Zabarin, Two-phase boundary layer with incompressible carrier phase on a plate with mass injection and suction on the surface. J. Appl. Mech. Tech. Phys. 54–61 (1987) (in Russian)

E.S. Asmolov, Motion of suspension in the laminar boundary layer on a flat plate. Fluid Dyn **27**, 49–54 (1992)

E.S. Asmolov, Dispersed phase motion in laminar boundary layer over a wedge. Fluid Dyn **28**, 778–784 (1993)

E.S. Asmolov, Dusty-gas flow in a laminar boundary layer over a blunt body. J Fluid Mech **305**, 29–46 (1995)

B.A. Balanin, V.A. Lashkov, Drag of a flat wedge in a two-phase flow. Fluid Dyn **17**, 317–321 (1982)

C.E. Brennen, *Fundamental of Multiphase Flow* (Cambridge University Press, Cambridge, 2006)

A.J. Chamka, J. Peddieson Jr, Singular behavior in boundary-layer flow of a dusty gas. AIAA J **30**, 2966–2968 (1992)

R. Clift, J.R. Grace, M.E. Weber, *Bubbles, Drops and Particles* (Academic Press, New York, 1978)

C.T. Crowe, M. Sommerfeld, T. Tsuji, *Fundamentals of Gas-Particle and Gas-Droplet Flows* (CRC Press, Boca Raton, 1998)

D.A. Drew, Mathematical Modeling of Two-Phase Flow. Ann Rev Fluid Mech **15**, 261–291 (1983)

S.K. Friedlander, *Smoke, Dusts and Haze: Fundamentals of Aerosol Behavior* (Wiley, New York, 1977)

A.M. Grishin, V.I. Zabarin, Heat transfer and friction in the two-phase boundary layer on a flat plate. J. Appl. Mech. Tech. Phys. 78–86 (1988) (in Russian)

M. Hussainov, A. Kartushinsky, A. Mulgi, U. Rudi, S. Tisler, Experimental and theoretical study of the distribution of mass concentration of solid particles in the two-phase laminar boundary layer on a flat plate. Int J Multiph Flow **21**, 1141–1161 (1995)

M. Ishii, *Thermo-Fluid Theory of Two-Phase Flows* (Eyrolles, Paris, 1975)

D. Kaftori, G. Hestroni, S. Banerjee, Particle behavior in the turbulent boundary layer II: Velocity and distribution profiles. Phys Fluids A **7**, 1107–1121 (1995)

A.I. Kartushinsky, I.A. Krupensky, S.V. Tisler, M.T. Hussainov, I.N. Shcheglov, Deposition of solid particles in laminar boundary layer on a flat plate. High Temp **47**, 92–900 (2009)

E. Loth, *Simulating Dynamics of Dispersed Multi-Phase Flow* (Cambridge University Press, Cambridge, 2006)

E.E. Michaelides, *Particles, Bubbles and Drops* (World Scientific Publishing, New York, 2006)

V.A. Naumov, Calculation of laminar non-isothermal boundary layer on a flat plate with account for lifting forces exerted of dispersed particles. High Temp **28**, 814–816 (1990). (in Russian)

R.I. Nigmatulin, *Dynamics of Multiphase Media* (Hemisphere, New York, 1991)

A.N. Osiptsov, Structure of a dispersed-mixture laminar boundary layer on a flat plate. Fluid Dyn **15**, 48–54 (1980). (in Russian)

A.N. Osiptsov, Mathematical modeling of dusty-gas boundary layers. ASME Appl Mech Rev **50**, 357–370 (1997)

A.N. Osiptsov, G.N. Shapiro, Two-phase flow over a surface with the formation of a liquid film by particles deposition. Fluid Dyn **24**, 559–566 (1989)

A.N. Osiptsov, G.N. Shapiro, Heat transfer in the boundary layer of a 'gas-evaporating drops' two-phase mixture. Int J Heat Mass Transf **36**, 71–78 (1993)

C.B. Rogers, J.K. Eaton, The behavior of small particles in a vertical turbulent boundary layer in air. Int J Multiph Flow **16**, 819–834 (1990)

C.B. Rogers, J.K. Eaton, The effect of small particles on fluid turbulence in a flat plate, turbulent boundary layer in air. Phys Fluids A **3**, 928–937 (1991)

M. Shin, D.S. Kim, J.W. Lee, Deposition of inertia-dominated particles inside a turbulent boundary layer. Int J Multiph Flow **29**, 93–926 (2003)

S.L. Soo, *Fluid Dynamics of Multiphase Systems* (Blaisdell Publishing Company, Waltham, 1967)

V.P. Stulov, Equations of laminar boundary layer in a two-phase medium. Fluid Dyn **14**, 37–44 (1979)

A. Taniere, B. Oesterle, J.C. Monnier, On the behaviour of solid particles in a horizontal boundary layer with turbulence and saltation effects. Exp Fluids **23**, 463–471 (1997)

AYu. Varaksin, *Turbulent Particles-Laden Gas Flows* (Springer, Berlin, 2007)

AYu. Varaksin, D.S. Mikhatulin, YuV Polezhaev, A.F. Polyakov, Measurements of velocity fields of gas and solid particles in the boundary layer of turbulized heterogeneous flow. High Temp **33**, 911–917 (1995)

J. Wang, E.K. Levy, Particle behavior in the turbulent boundary layer of a dilute gas-particle flow past a flat plate. Int J Exp Fluid Sci **30**, 473–483 (2006)

L.I. Zaichik, V.M. Alipchenkov, E.G. Sinaiski, *Particles in Turbulent Flows* (Wiley-VCH, Berlin, 2008)

# Chapter 2
# Laminar Mist Flows Over a Flat Plate with Evaporation

The experimental and numerical study of laminar two-phase boundary layer with evaporating liquid droplets performed in relatively small number of works (Bhatti and Savery 1975; Heyt and Larsen 1970; Hishida et al. 1980, 1982; Osiptsov and Shapiro 1989, 1993; Osiptsov and Korotkov 1998; Pakhomov and Terekhov 2012; Terekhov and Pakhomov 2002; Terekhov et al. 2000; Trela 1981). Nevertheless investigation of such flows is a topical problem for various short channels that are widely used in different heat transfer equipment.

Below there is a brief review of works devoted to investigation of hydrodynamics and heat and mass transfer in the laminar two-phase flows at the absence and or presence of evaporation on the surface of the particles.

Possibly the first numerical and experimental chapter in which studied the laminar boundary layer on dry isothermal surface at the gas-droplets flow process was performed in the work (Heyt and Larsen 1970). The system of differential equations in the approximation of the boundary layer includes the equation of continuity, impulse, energy and diffusion with the source members. The effect of droplets evaporation on the boundary layer structure, wall friction and heat transfer for the case of low droplets mass fraction of small particles ($M_{L1} < 0.05$) with high velocity of evaporation has been examined. It was shown that with the increase of water droplets mass concentration the increase of heat transfer is significant, whereas the friction growth on the wall is minor. Theoretical dependences for heat transfer and friction on the wall in the laminar gas-droplets flows have been obtained.

The chapter (Bhatti and Savery 1975) refers to the theory of the two-phase boundary layer on the plate at which the droplets penetrate into the boundary layer and evaporate without deposition on the plate surface. The equations of the droplets motion taking into account the drag forces, the lift force and the gravity have been obtained. With use of the theory is possible to calculate the upper and the lower limits of the particles size penetration in the boundary layer and evaporating time there without the surface wetting. Significant influence of the droplets diameter and their mass fraction on evaporation is shown. The increase of the dispersed phase content leads to the augmentation of the heat transfer between two-phase flow and the wall. Influence of the droplets initial diameter increase is

V. I. Terekhov and M. A. Pakhomov, *Flow and Heat and Mass Transfer in Laminar and Turbulent Mist Gas-Droplets Stream over a Flat Plate*, SpringerBriefs on Multiphase Flow, DOI: 10.1007/978-3-319-04453-8_2, © The Author(s) 2014

more complex: initial increase of the particles dimensions intensifies heat transfer and further with the growth of the droplets size heat transfer decreases.

The most detailed experimental studies of heat and mass transfer in the gas-drop flow streamlining the plane plate is the series of works of Hishida et al. (Hishida et al. 1980, 1982). In these works the heat transfer from the vertical flat plate to the laminar mist flow was studied. The average size of particles was $d_1 = 35$ µm. The other parameters of the two-phase flow: mass concentration of water $M_{L1} = 0–0.023$, the wall temperature is $t_W = 50–80$ °C, the initial temperature of air is $t_1 = 15–20$ °C and the Reynolds number built along the longitudinal coordinate and the velocity of the unperturbed flow is $Re_x = U_1x/v = (2.4–9.5) \times 10^4$.

Numerical study of dynamics and heat/mass transfer in a gas-droplet turbulent boundary layer on a vertical flat plate was carried out in the chapter (Pakhomov and Terekhov 2012). A large number of factors which affect the heat and mass transfer and the structure of thermal and concentration fields in a turbulent boundary layer is analyzed. It is shown that the increase in droplet concentration results in the intensification of heat transfer, as compared with the single-phase air flow.

A model for predicting heat and mass transfer in a laminar gas–vapor–droplets mist flow on a flat isothermal flat is developed in the chapter (Terekhov and Pakhomov 2002). Using this model, a numerical study is performed to examine the effect of thermal and flow parameters, e.g., Reynolds number, flow velocity, temperature ratio, concentration of the liquid phase, and drop size, on the profiles of velocity, temperature, composition of the two-phase mixture, and heat transfer intensification ratio. It is shown that, as the concentration of the liquid phase in the free flow increases, the rate of heat transfer between the plate surface and the vapor–gas mixture increases dramatically, whereas the wall friction increases insignificantly.

With the use of LDA measurements it was determined that the distribution of particles across the boundary layer is uniform and their velocity insignificantly exceeds the velocity of the air flow. The self-similarity velocity distribution in the two-phase flow does not depend on the mass flow rate ratio of water and air, wall temperature and velocity of the external flow and approximately agrees with the Blasius law (1955) for the laminar one-phase flow. In the wall region of the boundary layer the value of droplets velocity is somewhat higher than the Blasius profile. Authors of chapters (Hishida et al. 1980, 1982) explain this fact by the droplets inertia. The measurements have shown that sliding between the phases is practically absent in the considered range of the droplets sizes, the exception is only in the vicinity of the wall where the droplets insignificantly excel the gas phase. The averaged sliding of the phases according to the data of LDA was in the range of 0.03–0.07 m/s.

Authors derived that heat transfer intensification ratio at constant Reynolds number and the wall temperature was linearly depends on water mass flow rate ratio. It has been proved that intensification of heat transfer occurs for the account of the latent heat of liquid drops evaporation in the boundary layer. The value of heat transfer intensification is substantially affected by the value of the wall temperature. With the increase of the amount of water liquid the temperature of the air-vapor mixture and of the wall decreases and the length of evaporation area increases.

In the literature there are practically no data on the flow structure and influence of the droplets initial size on the heat transfer between the mist flow and the wall. The main aim of this work is to study the effect of the dispersed phase on the flow, structure of the boundary layer and heat transfer rate at thermal and gas dynamics parameters variation.

# References

M.S. Bhatti, C.W. Savery, Augmentation of heat transfer in laminar external boundary layer by vaporization of suspended droplets. Trans ASME J Heat Transf **97**, 179–184 (1975)

J.W. Heyt, P.S. Larsen, Heat transfer to binary mist flow. Int J Heat Mass Transf **14**, 1149–1158 (1970)

K. Hishida, M. Maeda, S. Ikai, Heat transfer from a flat plate in two-component mist flow. Trans ASME J Heat Transf **102**, 513–518 (1980)

K. Hishida, M. Maeda, S. Ikai, Heat transfer in two-component mist flow: boundary layer structure on an isothermal plate, vol. 4, in 7th International Heat Transfer Conference IHTC-7, Munich, Germany pp. 301–306 (1982)

A.N. Osiptsov, G.N. Shapiro, Two-phase flow over a surface with the formation of a liquid film by particles deposition. Fluid Dyn **24**, 559–566 (1989)

A.N. Osiptsov, G.N. Shapiro, Heat transfer in the boundary layer of a 'gas-evaporating drops' two-phase mixture. Int J Heat Mass Transf **36**, 71–78 (1993)

A.N. Osiptsov, D.V. Korotkov, Vapor-droplet boundary layer on the frontal surface of a hot blunted body. High Temp **36**, 75–281 (1998)

M.A. Pakhomov, V.I. Terekhov, Modeling of the flow structure and heat transfer in a gas-droplet turbulent boundary layer. Fluid Dyn **47**, 168–177 (2012)

H. Schlichting, *Boundary Layer Theory* (McGraw-Hill Publication House, New York, 1960)

V.I. Terekhov, M.A. Pakhomov, Numerical study of heat transfer in a laminar mist flow over an isothermal flat plate. Int J Heat Mass Transf **45**, 2077–2085 (2002)

V.I. Terekhov, M.A. Pakhomov, A.V. Chichindaev, Effect of evaporation of liquid droplets on the distribution of parameters in a two-species laminar flow. J Appl Mech Tech Phys **41**, 1020–1028 (2000)

M. Trela, An approximate calculation of heat transfer during flow of an air-water mist along heated flat plate. Int J Heat Mass Transf **24**, 749–755 (1981)

# Chapter 3
# Flow and Heat and Mass Transfer in a Laminar Gas-Droplets Boundary Layer

## 3.1 Physical Model

In the chapter is considered the steady flow of the gas-droplets flow in the laminar and turbulent boundary layer taking into account evaporation of monodispersed droplets, interfacial interaction, particles deposition on the plate surface, heat transfer deposited particles with the wall and vapor diffusion in the gas–vapor mixture.

Assume that the density of liquid drops substantially excels the density of the carrying agent (vapor or the vapor-air mixture). Because of the low volume content of the liquid phase ($\phi_1 \approx 10^{-4} \ll 1$) the inter-collision of the drops will not be taken into account (Crowe et al. 1998). The droplets are small (the initial diameter is $d_1 < 100$ μm), therefore in the flow their fragmentation and coalescence do not occur (Terekhov et al. 2005), since the Weber number based on the interfacial velocity $\mathrm{We} = \left|\vec{U} - \vec{U}_L\right|^2 \rho d / \sigma_L = 10^{-5} - 10^{-3}$, where $\sigma_L$ is a coefficient of the surface tension of liquid, is lesser that the critical Weber number according the data given in (Mastanaiah and Ganic 1981) $\mathrm{We}_C \approx 7$. The droplets keep their sphericity, i.e. Laplace number is $\mathrm{Lp} = \rho d \sigma_L / \mu^2 \gg 1$.

Note that the two-phase flows with monodispersed droplets do not exist in reality since in the dispersed medium there is always size distribution of the droplets, for example, log-normal function of distribution f(r). The most frequently applied is the normal logarithmic distribution law (Pazhi and Galustov 1984)

$$f\left(\frac{d}{2}\right) = \frac{1}{\sqrt{2}\ln(\sigma)} \cdot \exp\left[-\left(\frac{\ln(d/2A)}{\sqrt{2}\ln(\sigma)}\right)^2\right] \tag{3.1}$$

In the Eq. (3.1) $d$ is droplet diameter; $A$ is mathematical expectation and $\ln(\sigma)$ is mean squared deviation of the droplets radius logarithms. Nevertheless quite often the mean squared deviation of the drops dimensions does not exceed 8–10 % of some average value $d$. In this case the monodispersed model of the two-phase flow with sufficient accuracy describes the flow of real polydispersed medium.

V. I. Terekhov and M. A. Pakhomov, *Flow and Heat and Mass Transfer in Laminar and Turbulent Mist Gas-Droplets Stream over a Flat Plate*, SpringerBriefs on Multiphase Flow, DOI: 10.1007/978-3-319-04453-8_3,

The droplets deposited onto the wall from the two-phase flow instantly evaporate and the surface of the flat plate always remains dry that is quite fair for the heated walls. This is confirmed with experimental and numerical data for laminar and turbulent flow regimes (Mastanaiah and Ganic 1981; Pakhomov and Terekhov 2012; Terekhov and Pakhomov 2002; Terekhov et al. 2000; Yao and Rane 1980). This assumption is rather rough but is applicable at the conditions of this work since the difference between the wall and liquid phase temperatures is $t_W - t_L > 40\,°C$.

Conductive heat transfer by direct contact of the droplets with the surface is taken into account in the model (Mastanaiah and Ganic 1981; Pakhomov and Terekhov 2012; Terekhov and Pakhomov 2002; Terekhov et al. 2000; Yao and Rane 1980). Main mechanisms of heat transfer between the flow and the droplet are convection and thermal conductivity since in direct vicinity from the drop surface there is an immovable layer of the gas–vapor mixture via which heat is transferred only for the account of heat conductivity. In the work was used the two-stage mechanism of heat transfer: heat from the wall is transmitted to the flow and from the flow goes to liquid drops; the three-stage mechanism was considered as well (Mastanaiah and Ganic 1981):

(1)  heat from the wall is transferred to the droplets deposited onto the surface and is used up for their evaporation;
(2)  heat from the wall is transmitted to vapor–gas mixture;
(3a) part of heat from gas–vapor mixture is received by droplets and is used for their heat up and evaporation;
(3b) the remaining part of heat is used for the gas phase heating.

Radiation heat transfer was not considered in the chapter since according to the carried out estimation (Terekhov et al. 2000; Yao and Rane 1980) its share is not large (not more than 5 % of the total value of the heat flux).

All particles at the outer boundary of the boundary layer have the same size and their numerical concentration in the volume unit is also constant, at that, the latter condition is valid for the whole flow area. At complete evaporation of droplets they were replaced by pseudo-particles with the null diameter.

In the inflow the distribution of the gas–vapor mixture and droplets temperatures is uniform, at that, the carrier flow may be overheated in relation to saturation temperature at such pressure, and the temperature of the droplets surface is equal to the saturation temperature $T_S$. The temperature of the particle over its radius was also assumed constant since according to our assessment for the considered in the chapter range of initial conditions Bio number $Bi = \alpha d/\lambda_L$ lesser than 0.1, where $\alpha$ is the coefficient of heat transfer to the evaporating drop, $\lambda_L$ is the coefficient of thermal conductivity of liquid. In this case according to the results of (Lykov 1967) droplet temperature across its radius is constant. The role of diffusion may be neglected since within the limits of the boundary layer $\partial P/\partial r \approx 0$. There is insignificant influence of thermodiffusion as well.

## 3.2 The System of Governing Equations for Two-Phase Laminar Mixture

The heat and mass transfer of the gas-droplets flow in the laminar boundary layer is described by the set of steady-state equations of continuity, impulse in longitudinal direction, energy and diffusion for the vapor–gas mixture.

$$\frac{\partial U}{\partial x} + \frac{\partial V}{\partial y} = \frac{JnB}{\rho} \tag{3.2}$$

$$\rho\left(\frac{\partial U^2}{\partial x} + \frac{\partial (VU)}{\partial y}\right) = \frac{\partial}{\partial y}\left(\mu \frac{\partial U}{\partial y}\right) \tag{3.3}$$

$$\rho C_p \left(\frac{\partial (UT)}{\partial x} + \frac{\partial (VT)}{\partial y}\right) = \frac{\partial}{\partial y}\left(\frac{\mu}{\text{Pr}} \frac{\partial T}{\partial y}\right) - \alpha n B(T - T_L)$$
$$+ \rho D \frac{\partial K_V}{\partial y}(C_{PV} - C_{PA})\frac{\partial T}{\partial y} \tag{3.4}$$

$$\rho\left(\frac{\partial (UK_V)}{\partial x} + \frac{\partial (VK_V)}{\partial y}\right) = \frac{\partial}{\partial y}\left(\frac{\mu}{\text{Sc}} \frac{\partial K_V}{\partial y}\right) + JnB. \tag{3.5}$$

The system (3.2–3.5) are added by the equation of heat transfer on the interfacial boundary

$$\lambda_L\left(\frac{\partial T_L}{\partial y}\right)_d = \alpha(T - T_L) - JL \tag{3.6}$$

and the equation of vapor mass conservation on the evaporating surface of the droplet (Terekhov et al. 2000; Terekhov and Pakhomov 2002)

$$J = JK_V^* - \rho D \frac{\partial K_V^*}{\partial y}, \tag{3.7}$$

where $L$ is latent heat of evaporation; $K_V^*$ is vapor concentration on the surface of droplet surface, that agrees with the parameters of saturation at the temperature of the droplet $T_L$. The diffusion Stanton number $\text{St}_D$ have the form

$$\text{St}_D = -\rho D \frac{\partial K_V^*}{\partial r} \bigg/ \rho U(K_V^* - K_V), \tag{3.8}$$

The equation of vapor mass conservation (3.7) may be written as (Terekhov et al. 2000)

$$J = \text{St}_D \rho U b_{1D}, \tag{3.9}$$

where $b_{1D} = (K_V^* - K_V)/(1 - K_V^*)$ is the diffusion parameter of blowing determined with the use of the saturation curve.

The equation of heat and mass transfer from the surface of the non-evaporating sphere are given in (Clift et al. 1978; Soo 1967)

$$\mathrm{Nu}_P = \alpha_P d / \lambda = 2 + 0.6 \mathrm{Re}_L^{1/2} \mathrm{Pr}^{1/3} \quad \text{and} \quad \mathrm{Sh}_P = \beta d / D = 2 + 0.6 \mathrm{Re}_L^{1/2} \mathrm{Sc}^{1/3}.$$

Here $\mathrm{Re}_L = d \sqrt{(U - U_L)^2 + (V - V_L)^2} \Big/ v$ is Reynolds number of the particle based on the interfacial velocity; $\mathrm{Sh}_P$ is Sherwood number of the particle; $\beta$ is mass transfer coefficient. Diffusional Stanton number in Eq. (3.7) may be neglected according to the expression of (Kutateladze and Leont'ev 1989)

$$\mathrm{St}_D = \mathrm{Sh}_P / (\mathrm{Re}_L \mathrm{Sc}), \tag{3.10}$$

Then, the Eq. (3.9) with taking into account Eqs. (3.6–3.8) is transformed to the following form

$$J = \frac{\left(2 + 0.6\,\mathrm{Re}_L^{1/2}\mathrm{Sc}^{1/3}\right) b_{1D}}{\mathrm{Re}_L \mathrm{Sc}} \rho(\vec{U} - \vec{U}_L) \tag{3.11}$$

The coefficient of heat transfer for the evaporating droplet $\alpha$ is bound with the heat transfer coefficient of non-evaporating particles $\alpha_P$ (Mastanaiah and Ganic 1981)

$$\alpha = \frac{\alpha_p}{1 + C_p(T - T_L)/L} \tag{3.12}$$

The equation of material balance for the binary vapor–gas mixture takes the form

$$K_A + K_V = 1. \tag{3.13}$$

For the triple mixture vapor-gas-liquid it is written as follows

$$M_A + M_V + M_L = 1. \tag{3.14}$$

The relation between the values of mass concentrations of the mixture components $K$ and $M$ may be written as follows

$$K_V = M_V / (M_A + M_V); \quad K_A = M_A / (M_A + M_V) = 1 - K_V. \tag{3.15}$$

The expression for the calculation of the current diameter of the drop in $i$-calculated section is

$$d_i^3 = d_{i-1}^3 - J d_{i-1}^2 \frac{6\Delta x}{\rho_L U_{mi}} \tag{3.16}$$

## 3.3 Boundary Conditions

On the wall for the gas-phase velocity the conditions of no-slip and impermeability are performed

$$U = V = \frac{\partial K_V}{\partial r} = 0; \, T = T_W \tag{3.17}$$

At the outer edge of the boundary layer the following conditions were used.

$$U = U_1; V = V_1; T = T_1; M_L = M_{L1}; T_L = T_{L1}; d = d_1; K_V = K_{V1}. \tag{3.18}$$

The set of Eqs. (3.2–3.16) with boundary conditions (3.17) and (3.18) are the closed system of equations describing heat and mass transfer in the gas-droplets flow in the laminar boundary layer and allowing prediction next parameters: temperature distribution, phase concentration and component of vapor–gas mixture, as well as flow dynamics, particles size and analyze the heat transfer intensification ratio with droplets evaporation.

## 3.4 Numerical Realization

Numerical solution of differential equations was obtained with the use of the even finite-difference scheme of Crank-Nicolson presented in the work (Anderson et al. 1984), by the transformation of initial partial differential equations into the system of linear algebraic equations. The set of equations with tridiagonal matrix is solved by the sweep method with Thomas algorithm given in (Anderson et al. 1984).

The differential equation was considered in its general form.

$$Z\frac{\partial \Omega}{\partial x} = \frac{1}{r}\frac{\partial}{\partial r}\left(r\theta\frac{\partial \Omega}{\partial r}\right) + S, \tag{3.19}$$

where $Z$ are constant parameters; $\Omega$ is the differentiated parameter; $\theta$ is proportionality coefficient, $\theta$: $\lambda$, $D$; $S$ are terms of equation without the differentiated parameter. Now write the differential Eq. (3.19) in the finite-difference form according to Crank–Nicholson scheme (Anderson et al. 1984)

$$Z\frac{\Omega_{i+1,j} - \Omega_{i,j}}{\Delta x} = \frac{1}{y}\theta$$

$$\times \left(\frac{\Omega_{i+1,j} - \Omega_{i,j}}{\Delta y} + y\frac{\Omega_{i+1,j+1} + \Omega_{i,j+1} - 2(\Omega_{i+1,j} + \Omega_{i,j})}{2\Delta y^2} + \frac{\Omega_{i+1,j-1} + \Omega_{i,j-1}}{2\Delta y^2}\right) + S. \tag{3.20}$$

Here $\Delta x$, $\Delta y$ is the grid size along the longitudinal and radial coordinates.

After the grouping of terms in the finite-difference analogue of the differential Eq. (3.20) in the respective node $(i,j)$ of the finite-difference grid as a final option write as follows

$$A_{i,j}\Omega_{i,j-1} + B_{i,j}\Omega_{i,j} + C_{i,j}\Omega_{i,j+1} = D_{i,j}. \qquad (3.21)$$

The coefficients $A$, $B$, $C$ and $D$ in Eq. (3.21) are considered as known and are the coefficients of the tridiagonal matrix.

The number of nodes in the transverse and longitudinal directions was selected in order to provide the algorithm stability and calculations economy. All computations performed for the laminar flow were carried out on the grid containing 101 nodes in the longitudinal direction and 51 in the transverse one.

Since the equations of the system are nonlinear in each calculated point the additional iterations were organized. For all variables the double precision was used for decrease the numerical error related to truncation. The following conditions of convergence were assumed: $|Y_i - Y_{i-1}| < 10^{-5}$, where $Y$: $U$, $T$, $K_V$ and $d$. The predictions are finish when all criteria are satisfied.

The following parameters at the inlet varied in the work: mass fraction of air and droplets, its size and wall temperature.

## 3.5 Numerical Results and Discussion

All numerical computations were carried out for the mist laminar flow. The gas and droplets velocity in the external flow was constant and equal to $U_1 = U_{L1} = 10$ m/s; temperature of the mixture $T_1 = 293$ K and in the external flow it had in the saturated state $T_1 = T_{1S}$, at that, the mass vapor concentration in the external flow was $M_{V1} = 0.014$. In the predictions next parameters were varied—Reynolds number $\mathrm{Re}_x = U_1 x/v = 10^3$–$10^5$ (longitudinal coordinate); wall temperature $T_W = 323$–$473$ K; mass concentration of the droplets $M_{L1} = 0$–$0.05$ and its diameter in the external flow $d_1 = 0$–$100$ µm.

Main attention in the following series of calculations was paid to the influence of the liquid phase concentration on the distribution of velocities and temperatures in the boundary layer. The computed normalized profiles of the velocities and temperatures depending on the transverse coordinate are shown in he Fig. 3.1, where $\omega = U/U_1$ is the velocity profile (the dashed line); $\Theta = (T_W - T)/(T_W - T_1)$—temperature profile in the gas-droplets flow (solid line). The numerical results coincide with the Blasius profile (Schlichting 1955) at the absent liquid phase ($M_{L1} = 0$). Increase of liquid concentration results in the larger fullness of the velocity profile that is caused by intense evaporation predominantly in the wall region. Presence of the droplets in the flow has larger effect on temperature distribution, that finally shall lead to the more intense growth of heat transfer compared with friction. Thickness of the boundary layer practically does not change with increase of droplets mass fraction.

**Fig. 3.1** Dimensionless profiles of gas phase velocity (*dashed lines*) and temperature (*solid lines*) in the mist laminar boundary layer. $Re_x = 5 \times 10^4$; $U_1 = 10$ m/s; $T_{L1} = T_1 = T_S = 293$ K; $M_{V1} = 0.014$; $d_1 = 30$ μm; $T_W = 373$ K. $1$—$M_{L1} = 0$; $2$—$0.01$; $3$—$0.025$; $4$—$0.05$

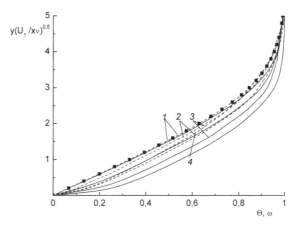

**Fig. 3.2** Droplets diameter profiles vs boundary layer cross-section. Solid curves are the data for droplets mass fraction $M_{L1} = 0.01$, dashed curves for $M_{L1} = 0.05$. $1$—$T_W = 473$ K; $2$—$373$; $3$—$323$

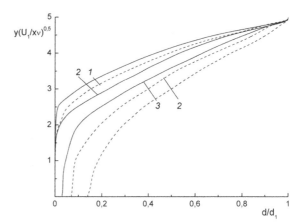

The abovementioned peculiarities in the distribution of velocity profiles and temperatures in the two-phase flow in the flat plate vertical boundary layer are manifested in the distribution of the particles size in the boundary layer cross-section. These data are given in the Fig. 3.2. The increase of the plate surface temperature caused the droplets evaporate faster, especially in the wall area and the zone of the one-phase gas–vapor flow increases. The growth of mass fraction of liquid phase (at constant diameter of the droplets) in the external flow the opposite trend is observed. The particles evaporate slower that can be explained by the increase of their numerical concentration and slower heat up of the boundary layer.

Profiles of the components of the mixture across the boundary layer shows a principally different view compared with the velocity and temperature profiles. Distributions of mass fraction of liquid, water vapor and air Results are shown in the Fig. 3.3. Since the plate surface is impermeable the derivative concentrations in the transverse direction of all substances on it are equal to null.

**Fig. 3.3** Mass fraction
distributions of the
components of the two-phase
mixture across the boundary
layer: **a** liquid phase; **b** vapor;
**c** air. The experimental
conditions are the same as for
Fig. 3.1. *Solid curve*
$M_{L1} = 0.01$; *dashed curve*
$M_{L1} = 0.05$. *1—*
$T_W = 323$ K; *2—373 K; 3—*
473 K

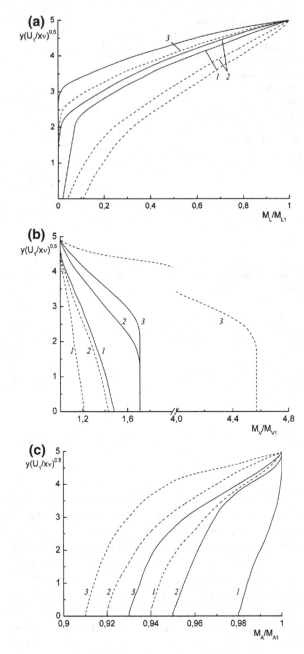

The mass concentration of liquid phase (Fig. 3.3a) decreases uninterruptedly
approaching to the wall for the account of evaporation processes. The wall area is
completely free of droplets for low values of liquid concentration $M_{L1}$ in the flow

external flow and high surface temperature. Here the region of one-phase air-vapor flow is formed. It can be obviously shown in the Fig. 3.3a, that with the decrease of droplets concentration in the external flow and with the growth of the wall temperature the area of one-phase flow appears earlier and its size increase. The increases of the wall temperature and liquid phase lead to growth of vapor mass fraction in the boundary layer. This is demonstrated in the Fig. 3.3b. The augmentation of vapor mass concentration is significant (in several times in comparison with the value in the free-stream flow).

For the considered in the chapter conditions when the external flow predominantly contains air ($M_{A1} \approx 0.93$–$0.98$), it should be expected that the relative change in the boundary layer will be small as well. This conclusion is proved by the results of simulations presented in the Fig. 3.3c. Maximum decrease of its content in the vicinity of the wall does not exceed 10 % in this case. The air mass concentration change more intensely over the boundary layer thickness with the growth of droplets mass fraction and wall temperature (see Fig. 3.3c).

All above listed peculiarities of the behavior of local parameters of gas–vapor flow effect on wall friction and heat transfer in the mist laminar flow. The presence of liquid phase exerts insignificant effect on surface friction coefficient (see Fig. 3.4). This agrees with the data of the Fig. 3.1. The predicted curve of influence of Reynolds number on the wall friction obtained at different droplets mass fraction have the same inclination angle and for one-phase flow ($M_{L1} = 0$) coincides with Blasius law (Schlichting 1960)

$$C_f \big/ 2 = 0.332 \big/ \mathrm{Re}_x^{1/2} \qquad (3.22)$$

The Fig. 3.5 shows effect of evaporation on the change of the wall friction coefficient $C_f$ for various droplets mass concentration. Friction increases insignificantly at the increase of droplets concentration and their initial diameter, where $C_{fA}$ is the wall friction coefficient in the one-phase air flow over the flat plate.

For heat transfer an absolutely different trend is observed (see Fig. 3.6). Presence of evaporating droplets in the flow leads to significant increase of heat transfer. The growth of mass concentration of the liquid phase leads abrupt rise of the curves of correlation $\mathrm{Nu}_x = f(\mathrm{Re}_x)$ and further down along the flat plate with expansion of the zone of one-phase flow their incline angle becomes closer to the ones in the case of one-phase laminar boundary layer (Schlichting 1960; Kutateladze and Leont'ev 1989)

$$\mathrm{Nu}_x = 0.332 \mathrm{Re}_x^{1/2} \mathrm{Pr}^{1/3} \qquad (3.23)$$

The local Nusselt number $\mathrm{Nu}_x$ at $T_W = \mathrm{const}$ based on the difference of the wall temperature and the temperature in the free-stream flow

$$\mathrm{Nu}_x = \frac{-(\partial T / \partial y)_W x}{(T_W - T_1)} \qquad (3.24)$$

**Fig. 3.4** Wall friction in the laminar gas–droplets flow. Points are calculated results for one-phase flow by the formula (3.22). $U_1 = 10$ m/s; $d_1 = 30$ μm; $T_W = 373$ K. $1$—$M_{L1} = 0$; $2$—$0.01$; $3$—$0.025$; $4$—$0.05$

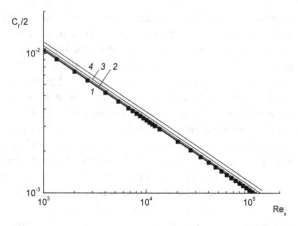

**Fig. 3.5** The effect of droplets mass fraction and its initial size on the wall friction coefficient. $\text{Re}_x = 5 \times 10^4$; $T = 373$ K. $1$—$d_1 = 100$ μm; $2$—$50$; $3$—$30$; $4$—$1$

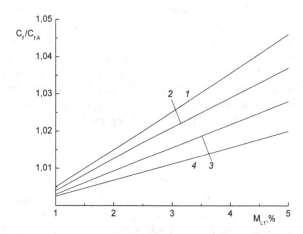

**Fig. 3.6** Heat transfer in the laminar gas–drop flow. Points are calculated results for one-phase flow by the formula (3.23) $0.1$—$M_{L1} = 0$; $2$—$0.01$; $3$—$0.05$

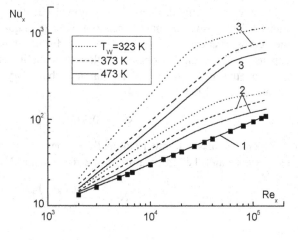

**Fig. 3.7** Effect of water droplets mass fraction on the heat transfer intensification ratio. $1$—Re$_x = 10^4$; $2$—$5 \times 10^4$; $3$—$10^5$

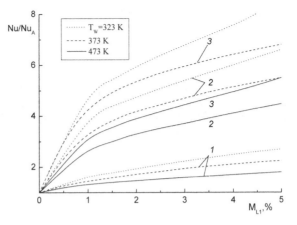

In Figs. 3.7, 3.8 and 3.9 the influence of thermal- and gasdynamics parameters of the two-phase gas-droplets flow on the of heat transfer intensification ratio Nu/Nu$_A$ is analyzed, where Nu$_A$ is Nusselt number at air flow in the one-phase air flow for identical initial conditions.

The heat transfer enhancement ratio may increase with the growth of droplets mass concentration up to 10 times in comparison with the one-phase air flow. However, as it is seen in the Fig. 3.7, the highest growth is observed in the area of small mass fraction of liquid phase ($M_{L1} < 0.01$–0.02). Further increase of droplets mass content results in insignificant growth of heat transfer, therefore, use of such flows for the purposes of heat transfer intensification is irrational.

The noticeable influence of wall temperature $T_W$ on the heat transfer intensification ratio is rendered. The temperature increase leads to faster droplets evaporation and therefore the wall region is freed from the liquid phase more rapidly and heat transfer intensity decreases.

One of the main parameters influenced on heat and mass transfer processes in the two-phase flows is the size of liquid droplets in the external flow. The results of this series of computations are demonstrated in the Fig. 3.8. The growth of the droplets diameter (at its mass fraction being constant) the area of the interfacial surface substantially decreases. Therefore, the rate of heat and mass transfer decreases. For small droplets ($d_1 < 10$ μm) the regime of equilibrium evaporation is realized and the intensification ratio for such range of droplets sizes has no dependence on the droplets diameter.

In the Fig. 3.9 the results on the influence of the wall temperature $T_W$ on the intensity of heat transfer in the gas-droplets flow is showed. The increase of $T_W$ leads to the decrease of the parameter of heat transfer enhancement due to more intense heating of the boundary layer and more rapid evaporation of the droplets. These conclusions qualitatively agree with the results of (Terekhov et al. 2005; Mastanaiah and Ganic 1981; Pakhomov and Terekhov 2010; Terekhov and Pakhomov 2003; Terekhov and Pakhomov 2009a, b) for the turbulent gas-droplets flow in a pipe.

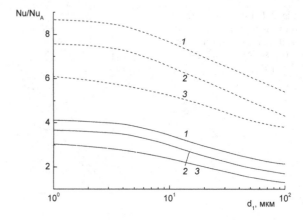

**Fig. 3.8** Effect of droplets diameter on heat transfer enhancement. $U_1 = 10$ m/s; $M_{L1} = 0.05$. *Solid lines* $Re_x = 10^4$; *dashed lines* $Re_x = 5 \times 10^4$. *1—* $T_W = 323$ K; 2—373 K; 4— 473 K

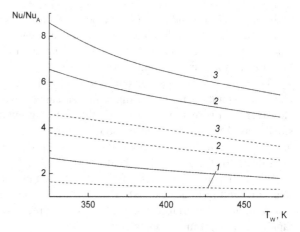

**Fig. 3.9** Effect of wall temperature on the heat transfer intensification ratio. $d_1 = 30$ μm; $U_0 = 10$ m/s. *Solid curves* $M_{L1} = 0.01$; *dashed curves* $M_{L1} = 0.05$. *1—*$Re_x = 10^4$; 2—$5 \times 10^4$; 3—$10^5$

Note one more peculiarity of the computational results in the Figs. 3.6, 3.7, 3.8 and 3.9. Increase of Reynolds number $Re_x$ (longitudinal coordinate) results in significant increase of heat transfer. The similar trend is observed for other regime conditions as well.

## 3.6 Comparison with Experimental Results

Results of computations are compared with the experimental data of (Hishida et al. 1980, 1982). The most detailed data are presented in the chapter (Hishida et al. 1980), where heat transfer from the flat vertical plate was investigated in the mist laminar flow. The mean diameter of water droplets in the external flow lied in the range 34–38 μm; mass concentration of liquid phase was $M_{L1} < 2.5$ %. The experimental conditions mostly agreed with physical and mathematical statement

**Fig. 3.10** Distributions of gas and droplets velocities across the flat plate boundary layer. *1*—Blasius's profile (Schlichting 1960); *2, 3*—measurements of (Hishida et al. 1980); *4*—predictions of our numerical model

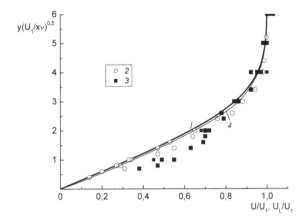

**Fig. 3.11** Comparison between our predicted results and the experimental data of (Hishida et al. 1980). Effect of Reynolds number and flow velocity on heat transfer. *Lines* are numerical results, points are experimental data by (Hishida et al. 1980). *1*—$U_1 = 9.8$ m/s; *2*—7.5; *3*—5.4

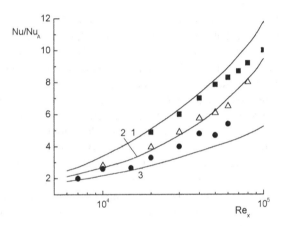

in the numerical model—the interfacial velocity equal zero value phase and droplets deposition on the wall is not take into account and wall surface was isothermal. All computational and experimental data were compared as heat transfer intensification ratio on thermogasdynamic parameters.

Results of comparison with experimental data of the work (Hishida et al. 1980) are given in the Fig. 3.10. The measured profiles of gas and droplets velocity well agree with the data of our simulations and fairly correlate with the Blasius profile (Schlichting 1955). Interfacial slip is practically absent in the range of initial diameter of the droplets considered in the work, the exception is observed only in the wall area where droplets slightly excel the gas phase velocity. For the case of fine-dispersed two-phase flow with small mass concentrations of the particles in first approach it is possible to use one-velocity model allowing significant simplification of the problem solution.

The distributions of heat transfer enhancement ratio $Nu/Nu_A$ on Reynolds number $Re_x$ at different flow velocities is presented in the Fig. 3.11. With the

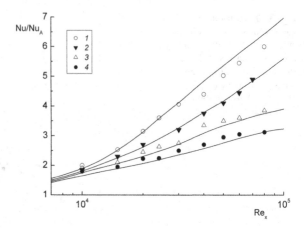

**Fig. 3.12** Heat transfer in the mist flow at various wall temperatures. *1—* $T_W = 323$ K; *2—333; 3—* *343; 4—353*

increase of Reynolds number heat transfer rate in the mist laminar flow increases as well that agrees with our computed data (see Figs. 3.6 and 3.7). At the same time the increase of the external flow velocity at fixed $Re_x$ exerts the augmentation of heat transfer. The increase of the flow velocity effects strong on the heat transfer intensification. The Reynolds number in this case is not the only determinative criterion and the processes of heat transfer are affected by the large number of thermal and gas dynamics parameters. This leads to the functional dependence of the local heat transfer in the gas-droplets flow both on the number $Re_x$, and the flow velocity or the longitudinal coordinate.

Influence of the heat transfer surface temperature on the heat transfer intensification ratio is demonstrated in the Fig. 3.12. Main conclusion inferred from the presented data proves significant heat transfer decrease with the increase of wall temperature. This decrease can not be explained by the decrease of the gas–vapor mixture density at the heat transfer surface that for the laminar flow is proportional $\psi = (T_W/T_1)^{-0.01}$ (Kutateladze and Leont'ev 1989). The main process in this physical phenomenon is blowing of vapor from the evaporating droplets. The increased temperature of the surface leads to fast formation of vapor film being further the buffer between the wall and the two-phase flow. Such heat transfer behavior at the variance of the temperature factor is pointed out in many experimental and numerical investigations (Terekhov and Pakhomov 2002; Terekhov et al. 2000; Hishida et al. 1980, 1982; Heyt and Larsen 1970).

Analysis of the influence of mass concentration of liquid particles on the local heat transfer carried out with the use of the data in the Fig. 3.13 for two temperatures of the wall $T_W = 323$ K (a) and 353 K (b). The significant effect of heat transfer enhancement is observed even at relatively small contents of liquid. The heat transfer increases 5–7 times for mass fraction of water droplets $M_{L1} \approx 2.5$ %.

The comparison of numerical results with the experimental data (Heyt and Larsen 1970) is given in the Fig. 3.14. Effects of heat transfer intensification for the account of evaporation processes turned out to be decreased compared with the experiments of (Heyt and Larsen 1970) and predictions of this work.

**Fig. 3.13** The effect of droplets mass fraction on the heat transfer enhancement ratio for various wall temperature $T_W = 323$ K (**a**) и 353 K (**b**). *1*— $U_1 = 9.8$ m/s; *2*—5.4

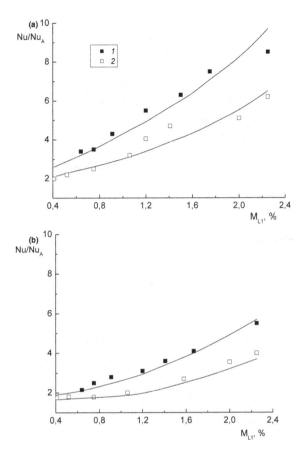

**Fig. 3.14** Comparison between the predicted results with the experimental and theoretical data of (Heyt and Larsen 1970). *Lines* are numerical results: *solid curves* are our results, *dashed curves* are predictions of (Heyt and Larsen 1970) and points are experimental data by (Heyt and Larsen 1970). *1*—$M_{L1} = 0$; *2* 0.023; *3* calculations by the formula (3.23)

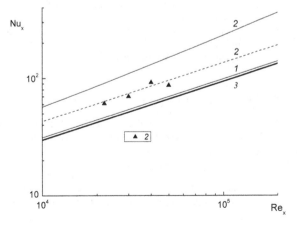

The dependence of $Nu_x$ number on $Re_x$ number for the two-phase regime is situated equidistant to the law of heat transfer in the one-phase flow (Schlichting 1960; Kutateladze and Leont'ev 1989). There is no strong explanation of such behavior of the heat transfer results (Heyt and Larsen 1970) in gas-droplets flow. There may be rather many possible reasons that cause under prediction of heat transfer augmentation, such as polydispersity, droplets deposition, etc.

The general conclusion deduced from Figs. 3.10, 3.11, 3.12, 3.13 and 3.14 is as follows. The computational results of the chapter correctly describe the experimental data set. The physical model takes into account main thermogasdynamic processes that occur in the two-phase gas-drop boundary layer despite the large number of assumptions.

# References

C.T. Crowe, M. Sommerfeld, T. Tsuji, *Fundamentals of Gas-Particle and Gas-Droplet Flows* (CRC Press, Boca Raton, 1998)

V.I. Terekhov, M.A. Pakhomov, K.A. Sharov, N.E. Shishkin, The thermal efficiency of near-wall gas-droplets screens. II. Experimental study and comparison with numerical results. Int J Heat Mass Transf **48**, 1760–1771 (2005)

K. Mastanaiah, E.N. Ganic, Heat transfer in two-component dispersed flow. Trans ASME J Heat Transf **103**, 300–306 (1981)

D.G. Pazhi, V.S. Galustov, *Basics of atomization techniques* (Chemistry Publication House, Moscow, 1984). (in Russian)

M.A. Pakhomov, V.I. Terekhov, Modeling of the flow structure and heat transfer in a gas-droplet turbulent boundary layer. Fluid Dyn **47**, 168–177 (2012)

V.I. Terekhov, M.A. Pakhomov, Numerical study of heat transfer in a laminar mist flow over an isothermal flat plate. Int J Heat Mass Transf **45**, 2077–2085 (2002)

V.I. Terekhov, M.A. Pakhomov, A.V. Chichindaev, Effect of evaporation of liquid droplets on the distribution of parameters in a two-species laminar flow. J Appl Mech Tech Phys **41**, 1020–1028 (2000)

S-Ch. Yao, A.G. Rane, Heat transfer of laminar mist flow in tubes. Trans ASME J Heat Transf **102**, 678–683 (1980)

A.V. Lykov, *Theory of Heat Conductivity* (The Highest School Publication House, Moscow, 1967). (in Russian)

R. Clift, J.R. Grace, M.E. Weber, *Bubbles, Drops and Particles* (Academic Press, New York, 1978)

S.L. Soo, *Fluid dynamics of multiphase systems* (Blaisdell Publication Company, Waltham, 1967)

D.A. Anderson, J.C. Tannehill, R.H. Pletcher, *Computational Fluid Mechanics and Heat Transfer* (Hemisphere, New York, 1984)

H. Schlichting, *Boundary Layer Theory* (McGraw-Hill Publication House, New York, 1960)

S.S. Kutateladze, A.I. Leont'ev, *Heat and Mass Transfer in Turbulent Boundary Layer* (Hemisphere, New York, 1989)

M.A. Pakhomov, V.I. Terekhov, Enhancement of an impingement heat transfer between turbulent mist jet and flat surface. Int J Heat Mass Transf **53**, 3156–3165 (2010)

V.I. Terekhov, M.A. Pakhomov, Numerical simulation of hydrodynamics and convective heat transfer in turbulent tube mist flow. Int. J. Heat Mass Transfer **46**, 1503–1517 (2003)

V.I. Terekhov, M.A. Pakhomov, Film-cooling enhancement of the mist vertical wall jet on the cylindrical channel surface with heat transfer. Trans. ASME J. Heat Transf. 131:062201 (2009a)

V.I. Terekhov, M.A. Pakhomov, Predictions of turbulent flow and heat transfer in gas-droplets flow downstream of a sudden pipe expansion. Int J Heat Mass Transf **52**, 4711–4721 (2009b)

K. Hishida, M. Maeda, S. Ikai, Heat transfer from a flat plate in two-component mist flow. Trans ASME J Heat Transf **102**, 513–518 (1980)

K. Hishida, M. Maeda, S. Ikai, Heat transfer in two-component mist flow: boundary layer structure on an isothermal plate, vol. 4, in *7th International Heat Transfer Conference IHTC-7*, Munich, Germany, pp. 301–306 (1982)

M.A. Pakhomov, M.V. Protasov, V.I. Terekhov, AYu. Varaksin, Experimental and numerical investigation of downward gas-dispersed turbulent pipe flow. Int J Heat Mass Transf **50**, 2107–2116 (2007)

J.W. Heyt, P.S. Larsen, Heat transfer to binary mist flow. Int J Heat Mass Transf **14**, 1149–1158 (1970)

# Chapter 4
# Numerical Modeling of Flow and Heat Transfer in a Turbulent Gas-Droplets Boundary Layer

## 4.1 Eulerian Numerical Model for the Two-Phase Flow

Wall-bounded turbulent gas-droplets flows with phase changes employ very often in a vide range of industrial processes. The addition of the particulate phase to the already complicated turbulent flow substantially increased the description of the problem. A better understanding of the physics of such flows would make it possible to develop improved equipment design, longer components life, reduced maintenance costs, increased efficiency, and other operational improvements.

Over the past 10 years multiphase computational fluid dynamics is increasingly used by industry for process analysis and optimization. Most of these processes involve multi-phase flows with liquid or solid particles in complex geometries which may be also accompanied by heat and mass transfer. Numerical computations of multiphase flows may be performed on different levels of complexity related to the resolution of the interface between the phases and the turbulence modeling. For engineering problems two approaches based on the Reynolds-averaged Navier–Stokes equations (RANS) are commonly applied, namely the two-fluid or Euler/Euler approach and the Euler/Lagrange method. In order to account for the interaction between phases, i.e. momentum exchange and heat and mass transfer, the conservation equations have to be extended by appropriate source/sink terms.

In the two-fluid Eulerian approach both phases are considered as interacting continua. Properties such as the mass of particles per unit volume are considered as a continuous properly and the particle velocity is the averaged velocity over an averaging control volume or computational cell. Also the interfacial transfer of mass, momentum, or energy requires averaging over the computational cells. In the Eulerian formulation, the dispersed solid phase is treated as a second continuous fluid. The governing equations of particle motion have a similar differential form to fluid flow equations and separate boundary conditions are applied to each phase. The most noteworthy advantage of the Eulerian scheme is that the solution of the equations of motion yields average flow conditions, such as particle concentration and velocity, in each computational cell. Hence, the Eulerian model is

V. I. Terekhov and M. A. Pakhomov, *Flow and Heat and Mass Transfer in Laminar and Turbulent Mist Gas-Droplets Stream over a Flat Plate*, SpringerBriefs on Multiphase Flow, DOI: 10.1007/978-3-319-04453-8_4,

computationally more efficient compared with its Lagrangian counterpart. The interaction between phases (coupling) is easily considered in Eulerian models by the addition of extra terms in the relevant equations. The numerical procedures utilized to solve the fluid phase equations can also be applied to the dispersed phase.

The Euler/Lagrange approach is only applicable to dispersed two-phase flows and accounts for the discrete nature of the individual particles. The dispersed phase is modelled by tracking a large number of particles through the flow field in solving the equations of motion taking into account the relevant forces acting on the particle. Generally, the particles are considered as point-particles, i.e. the finite dimension of the particles is not considered and the flow around the individual particles is not resolved. Since the number of real particles in a flow system is usually too large for allowing a tracking of all particles, the trajectories of computational particles (i.e. parcels) which represent a number of real particles with the same properties (i.e. size, velocity and temperature) are calculated. In stationary flows a sequential tracking of the parcels may be adopted, while in unsteady flows all parcels need to be tracked simultaneously on the same time level. Local average properties such as dispersed phase density and velocity are obtained by ensemble averaging. Statistically reliable results for each computational cell require the tracking of typically between $10^4$ and $10^5$ parcels, depending on the considered flow. The advantage of this method is that physical effects influencing the particle motion, such as particle-turbulence interaction, particle–wall collisions, and collisions between particles can be modelled on the basis of physical principles. Moreover, a particle size distribution may be easily considered by sampling the size of the injected particles from a given distribution function. Problems however may be encountered in the convergence behavior for high particle concentration due to the influence of the dispersed phase on the fluid flow (i.e. two-way coupling) which is accounted for by source terms obtained through averaging particle trajectories (Clift et al. 1978; Varaksin 2007).

An important question for the Eulerian–Lagrangian approach is calculation of dispersed phase concentration. For this purpose the procedures of spatial or time spatial (Crowe et al. 1998; Drew 1983) results averaging are applied for all particle trajectories within the control volume of the Eulerian net. However, these methods do not allow calculation of particle concentration with sufficient accuracy. Another approach to calculation of dispersed phase parameters uses the so-called full Lagrangian method by (Osiptsov 2000). Osiptsov's method is rooted in Lagrangian theory, the particle concentration being obtained from the Lagrangian form of the mass conservation equation by computing the change in volume of an element of 'particle fluid' along its pathline. The droplets mass fraction is then obtained algebraically from the continuity equation in the Lagrangian form. This method is based on application of additional equations for the components of Jacobian transition from Eulerian to Lagrangian variables. The main advantage of this approach is the fact that all parameters of dispersed phase, including concentration, are determined from the solutions to the systems of ordinary differential equations on the chosen trajectories of particles. The discussion and description of boundary conditions for Jacobian is presented in paper (Healy and Young 2005).

Essential for a reliable application of both methods is the appropriate modelling of the relevant physical mechanisms affecting the particle motion, as for example, turbulent transport of particles, wall interactions of particles, collisions between particles and agglomeration. In some cases the physical phenomena are far too complicated to allow for a derivation of the model from basic principles of physics (e.g. particle agglomeration). Therefore, detailed experiments are required to analyze the considered phenomenon and to derive appropriate empirical or semi-empirical models. In order to validate the models, the results of the numerical predictions need to be compared with bench mark test cases featuring the considered phenomenon.

In the present study we used the Euler approach (Drew 1983; Ishii 1975) implemented using the method of (Derevich 2002; Derevich and Zaichik 1988). The model is based on using PDF kinetic equations for the coordinates, velocities and temperatures of particles. According to this approach, in passing from the equations of instantaneous motion and heat transfer of a single particle to equations for an ensemble of dispersed particles, a particle distribution function (PDF), dependent on the coordinates $\vec{x}_L$, velocities $\vec{U}_L$ and temperature $T_L$, is used. It should be noted that this method was initially developed in (Derevich 2002; Derevich and Zaichik 1988) for treating two-phase flows laden with solid particles; we, however, subsequently showed that this method could also be successfully used for modeling flows with (Pakhomov and Terekhov 2010, 2012; Terekhov and Pakhomov 2002, 2003, 2009a, b; Terekhov et al. 2000, 2005) and without (Pakhomov et al. 2007) evaporation of droplets. In turbulent flow mode, it was assumed that, as a result of intense gas velocity pulsations in radial direction, the distribution of droplet sizes flattens over the boundary-layer cross-section.

## 4.2 Governing Equations for the Gas Phase

The numerical model based on the Eulerian/Eulerian approach. For the gas phase we used the set of steady-state Reynolds averaged Navier–Stokes (RANS) equations for the two-phase flow in boundary layer approach (Pakhomov and Terekhov 2012)

$$\rho \frac{\partial U_j}{\partial x_j} = \frac{6J\Phi}{d} \tag{4.1}$$

$$\rho \frac{\partial (U_i U_j)}{\partial x_i} = -\frac{\partial (P + 2k/3)}{\partial x_i} + \frac{\partial}{\partial x_j} \left[ (\mu + \mu_T) \left( \frac{\partial U_i}{\partial x_j} + \frac{\partial U_j}{\partial x_i} \right) \right]$$
$$- (U_i - U_{Li}) \frac{\Phi}{d} \times \left[ \frac{1}{8} C_D \rho |\vec{U} - \vec{U}_L| + J \right] + \rho_L g_u \langle u_i u_j \rangle \frac{\partial \Phi}{\partial x_j} \tag{4.2}$$

$$\rho \frac{\partial (U_i T)}{\partial x_i} = \frac{\partial}{\partial x_i} \left( \frac{\mu}{\mathrm{Pr}} + \frac{\mu_T}{\mathrm{Pr}_T} \right) \frac{\partial T}{\partial x_i} - \frac{6\Phi}{C_P d} [\alpha (T - T_L) + JL]$$
$$+ \frac{\rho D_T}{C_P} (C_{PV} - C_{PA}) \left( \frac{\partial K_V}{\partial x_i} \frac{\partial T}{\partial x_i} \right) + \frac{C_{PL} \rho_L \tau g_{ut}}{C_P} \langle u_j t \rangle \frac{\partial \Phi}{\partial x_j} \qquad (4.3)$$

$$\rho \frac{\partial (U_i K_V)}{\partial x_i} = \frac{\partial}{\partial x_i} \left( \frac{\mu}{\mathrm{Sc}} + \frac{\mu_T}{\mathrm{Sc}_T} \right) \frac{\partial K_V}{\partial x_i} + \frac{6J\Phi}{d} \qquad (4.4)$$

$$\rho = P/(\bar{R} T). \qquad (4.5)$$

Here $\Phi$ is the volume fraction of droplets, are coefficients of droplets entrainment into a large-eddy fluctuating motion of the gas phase

$$f_u = 1 - \exp\left(-\Omega^{\varepsilon L}/\tau\right), \quad f_{ut} = 1 - \exp\left(-\Omega^{tL}/\tau_\Theta\right),$$

$$g_u = \Omega^{\varepsilon L}/\tau - 1 + \exp\left(-\Omega^{\varepsilon L}/\tau\right). \qquad (4.6)$$

Here $\Omega^{\varepsilon L}$ is the interaction time between the droplet and turbulent eddy (Derevich 2002; Zaichik et al. 2008).

Value of turbulent Prandtl and Schmidt numbers were equal to $\mathrm{Pr}_T = \mathrm{Sc}_T = 0.85$. In addition we used correlation for Prandtl number from the work (Derevich 2002). Discrepancies in the results of Nusselt number calculations for one-phase flow at the use of $\mathrm{Pr}_T = 0.85$ and equation of the work (Derevich 2002) appeared to be insignificant (not more than 3 %).

The turbulent Reynolds stress and the turbulent heat and diffusion fluxes in the gas phase are determined using the Boussinesq hypothesis and have the form

$$\langle u_i u_j \rangle = -v_T \left( \frac{\partial U_i}{\partial x_j} + \frac{\partial U_j}{\partial x_i} \right) + \frac{2}{3} k \delta_{ij}, \ \langle t u_j \rangle = - \frac{v_T}{\mathrm{Pr}_T} \frac{\partial T}{\partial x_j},$$

$$\langle u_j k_V \rangle = - \frac{v_T}{\mathrm{Sc}_T} \frac{\partial K_V}{\partial x_j}, \qquad (4.7)$$

where in Eqs. (4.7) $v_T$ is the turbulent kinematic viscosity; $\delta_{ij}$ is Kronecker delta.

## 4.3  Two-Equations Turbulence Model

The turbulence two-equation model of (Hwang and Lin 1998) is designed not only to conform with the near-wall characteristics obtained with DNS data but also to posses the correct prediction of asymptotic behavior in the vicinity of the solid surface. The key features of this model are the adoption of the Taylor microscale in the damping function and the inclusion of the pressure diffusion terms in the $k$ and $\tilde{\varepsilon}$ equations. Those terms were ignored in many $k$–$\varepsilon$ models such as, for instance, the models of (Jones and Launder 1973; Myong and Kasagi 1990).

The equations for the turbulence kinetic energy $k$ and the dissipation rate of this energy $\tilde{\varepsilon}$ modified so that to take the presence of the dispersed phase into account have the form

$$\rho \frac{\partial (U_j k)}{\partial x_j} = \frac{\partial}{\partial x_j} \left[ \left( \mu + \frac{\mu_T}{\sigma_k} \right) \frac{\partial k}{\partial x_j} \right] - \frac{1}{2} \frac{\partial}{\partial x_j} \left[ \mu \frac{k}{\varepsilon} \frac{\partial \hat{\varepsilon}}{\partial x_j} \right] + \rho \Pi - \rho \varepsilon + S_k \qquad (4.8)$$

$$\rho \frac{\partial (U_j \tilde{\varepsilon})}{\partial x_j} = \frac{\partial}{\partial x_j} \left[ \left( \mu + \frac{\mu_T}{\sigma_\varepsilon} \right) \frac{\partial \tilde{\varepsilon}}{\partial x_j} \right] - \frac{\partial}{\partial x_j} \left( \mu \frac{\tilde{\varepsilon}}{k} \frac{\partial k}{\partial x_j} \right)$$
$$+ \frac{\rho \tilde{\varepsilon}}{k} (C_{\varepsilon 1} f_1 \Pi - C_{\varepsilon 2} \tilde{\varepsilon} f_2) + S_\varepsilon \qquad (4.9)$$

Here, $\mu_T = C_\mu f_\mu \rho k^2 / \tilde{\varepsilon}$ is the eddy dynamic viscosity, $\Pi = -\langle u_i u_j \rangle \frac{\partial U_i}{\partial x_j}$ is the production of turbulent kinetic energy, $\varepsilon = \tilde{\varepsilon} + \hat{\varepsilon}$ is the dissipation rate; $\hat{\varepsilon} = 2\mu/\rho$ $\left[ \partial (\sqrt{k})/\partial r \right]^2$ is the dissipation rate in the wall region ($y_+ \leq 15$); where $\hat{\varepsilon} = 0$ for $y_+ > 15$

Constants and damping functions have the form of the paper (Hwang and Lin 1998): $C_\mu = 0.09$; $\sigma_k = 1.4 - 1.1 \exp [-(0.1 y_\lambda)]$; $C_{\varepsilon 1} = 1.44$; $C_{\varepsilon 2} = 1.92$; $f_1 = f_2 = 1$; $\sigma_\varepsilon = 1.3 - \exp [-(0.1 y_\lambda)]$; $f_\mu = 1 - \exp (-0.01 y_\lambda - 0.008 y_\lambda^3)$; $\Omega^\varepsilon = (15 v/\varepsilon)^{1/2}$ is the Euler turbulent microscale (Zaichik 1999; Zaichik et al. 2008), $\tau = \rho_L d^2/(18 \mu W)$ is the relaxation time of dispersed phase with taking into account Stokes law, $W = (1 + \text{Re}_L^{2/3}/6)$.

The terms $S_k$ and $S_\varepsilon$ in Eqs. (4.8)–(4.9) which considered the extra dissipation of carrier phase turbulence energy by addition of fine droplets and the transfer of gas TKE with the averaged motion due to the mean inter-phase slipping in the flow with nonuniformly distributed droplets have the form (Zaichik 1999)

$$S_k = \frac{2 \rho_L \Phi k}{\tau} (1 - f_u) + \rho_L g_u \langle u_i u_j \rangle (U_i - U_{Li}) \frac{\partial \Phi}{\partial x_j}; \qquad (4.10)$$

$$S_\varepsilon = \frac{2 \rho_L \varepsilon}{\tau} \left[ \Phi (1 - f_\varepsilon) + \frac{\tau g_\varepsilon}{3} (U_i - U_{Li}) \frac{\partial \Phi}{\partial x_j} \right]. \qquad (4.11)$$

Here in Eqs. (4.10)–(4.11), $f_\varepsilon = 1 - \exp (-\Omega^\varepsilon/\tau)$, $g_\varepsilon = \Omega^\varepsilon/\tau - f_\varepsilon$ are coefficients of droplet entrainment into the micropulsational motion of the gas flow.

## 4.4 Dispersed Phase

The system of mean equations for the transport process in dispersed phase in the boundary layer approach has the form (Pakhomov and Terekhov 2010, 2012)

$$\frac{\partial (\rho_L U_{Lj})}{\partial x_j} = -\frac{6 J \Phi}{d} \qquad (4.12)$$

$$\frac{\partial\left(\rho_L \Phi U_{Lj} U_{Li}\right)}{\partial x_j} + \frac{\partial\left(\rho_L \Phi \langle u_{Li} u_{Lj}\rangle\right)}{\partial x_j} = \frac{\rho_L}{\tau}\left[\Phi(U_i - U_{Li}) - \frac{\partial(D_{Lij}\Phi)}{\partial x_j}\right] + \Phi g \quad (4.13)$$

$$\rho_L \frac{\partial(\Phi U_{Lj} T_{Li})}{\partial x_j} + \rho_L \frac{\partial}{\partial x_j}\left(\Phi\langle\theta u_{Lj}\rangle\right) = \Phi(T_i - T_{Li})\frac{\rho_L}{\tau_\Theta} - \frac{\rho_L}{\tau_\Theta}\frac{\partial\left(D^\Theta_{Lj}\Phi\right)}{\partial x_j}. \quad (4.14)$$

Here, $D_{Lij} = \tau(\langle u_{Li} u_{Lj}\rangle + g_u \langle u_i u_j\rangle)$, $D^\Theta_{Lij} = \tau_\Theta\langle t_L u_{Lj}\rangle + \tau g^\Theta_u \langle t u_j\rangle$ are tensors of turbulent diffusion and turbulent heat flux of particles, respectively (Zaichik 1999).

In the second Eq. of the system (4.12)–(4.14) we take into account the effect of droplets motion of the convective transport, turbulent transport (turbophoresys), aerodynamics drag, gravity, and turbulent diffusion. The main reason of appearance of the turbophoresys force is the gradient of velocity fluctuations of carrier phase. This force has directed towards to the decrease of the level of gas turbulence.

In the model it is not take into account the effect of Saffman (shear lift force) force on the dynamics of the dispersed phase. For the gas-dispersed turbulent flows the importance of taking into account the shear lift force is shown for in many papers, see e.g. (Chen 2000; Wang and Levy 2006; Marchioli et al. 2007).

The model for the transport of Reynolds stresses $\langle u_{Li} u_{Lj}\rangle$ and temperature fluctuations $\langle\theta^2_L\rangle$, and also the turbulent heat flux $\langle\theta_L u_{Lj}\rangle$ in dispersed phase were written using the equations by (Zaichik 1999)

$$\underbrace{U_{Lk}\frac{\partial\langle u_{Li} u_{Lj}\rangle}{\partial x_k}}_{I} + \underbrace{\frac{1}{\Phi}\left\{\frac{\partial}{\partial x_k}\left(\Phi\langle u_{Li} u_{Lj} u_{Lk}\rangle\right)\right\}}_{II} + \underbrace{\langle u_{Li} u_{Lk}\rangle\frac{\partial U_{Lj}}{\partial x_k} + \langle u_{Lj} u_{Lk}\rangle\frac{\partial U_{Li}}{\partial x_k}}_{III}$$

$$= \underbrace{\frac{2}{\tau}\left(f_u\langle u_i u_j\rangle - \langle u_{Li} u_{Lj}\rangle\right)}_{IV}$$

$$(4.15)$$

$$\underbrace{U_{Lk}\frac{\partial\langle\theta^2_L\rangle}{\partial x_k}}_{I} + \underbrace{\frac{1}{\Phi}\left\{\frac{\partial}{\partial x_k}\left(\Phi\langle\theta^2_L u_{Lk}\rangle\right)\right\}}_{II} + \underbrace{2\langle\theta_L u_{Lk}\rangle\frac{\partial T_L}{\partial x_k}}_{III} = \underbrace{\frac{2}{\tau_\Theta}\left(f_{ut}\langle t^2\rangle - \langle\theta^2_L\rangle\right)}_{IV} \quad (4.16)$$

$$\underbrace{U_{Lk}\frac{\partial\langle\theta_L u_{Lj}\rangle}{\partial x_k}}_{I} + \underbrace{\frac{1}{\Phi}\left\{\frac{\partial}{\partial x_k}\left(\Phi\langle\theta_L u_{Li} u_{Lk}\rangle\right)\right\}}_{II} + \underbrace{\langle u_{Li} u_{Lk}\rangle\frac{\partial T_L}{\partial x_k} + \langle\theta_L u_{Lk}\rangle\frac{\partial U_{Li}}{\partial x_k}}_{III}$$

$$= \underbrace{\left(\frac{f_{\Theta u}}{\tau} + \frac{f_{u\Theta}}{\tau_\Theta}\right)\langle u_i t\rangle - \left(\frac{1}{\tau} + \frac{1}{\tau_\Theta}\right)^{-1}\langle\theta_L u_{Lj}\rangle}_{IV}. \quad (4.17)$$

Here, $\Omega^{\varepsilon L}$ is the interaction time between the droplet and eddy (Derevich 2002)

$$\Omega^{\varepsilon L} = \begin{cases} \Omega^E, & |\vec{U} - \vec{U}_L|\Omega^E \leq \Gamma^E \\ \Gamma^E/|\vec{U} - \vec{U}_L|, & |\vec{U} - \vec{U}_L|\Omega^E > \Gamma^E \end{cases}, \tag{4.18}$$

where in the Eq. (4.18) $\Gamma^E = 2(\langle u^2 \rangle)^{1/2} \cdot \Omega^{Lag}$ integral length scale of the turbulent eddy; $\Omega^{Lag} = 0.608\Omega^E$ and $\Omega^E = 0.22\,k/\tilde{\varepsilon}$ are Lagrangian and Eulerian integral time scales (Derevich 2002); $\Omega^{tL}$ is interaction time between the droplets and thermal eddy. We assumed that $\Omega^{tL} \approx \Omega^{\varepsilon L}$ (see Terekhov et al. 2005).

In system (4.15)–(4.17), convective transport of fluctuations and temperature (I), diffusion (II), production of fluctuations from gradients of dispersed phase average motion and temperature (III), and the interaction between the phases (IV) are taken into account. Note that in the work two components of velocity fluctuations and Reynolds stresses of disperse phase were predicted.

For determination of diffusion component in system (4.15)–(4.17) the algebraic equation for the 3rd moment was used (Zaichik 1999).

$$\langle u_{Li}u_{Lj}u_{Lk}\rangle = -\frac{1}{3}\left(D_{Lkn}\frac{\partial\langle u_{Li}u_{Lj}\rangle}{\partial x_n} + D_{Ljn}\frac{\partial\langle u_{Li}u_{Lk}\rangle}{\partial x_n} + D_{Lin}\frac{\partial\langle u_{Lj}u_{Lk}\rangle}{\partial x_n}\right). \tag{4.19}$$

$$\langle \theta_L^2 u_{Lk}\rangle = -\frac{1}{\tau + 2\tau_\Theta}\cdot\left(\tau_\Theta D_{Lik}\frac{\partial\langle\theta_L u_{Lj}\rangle}{\partial x_k} + \tau_\Theta D_{Ljk}\frac{\partial\langle\theta_L u_{Li}\rangle}{\partial x_k} + \tau D_{Lk}\frac{\partial\langle u_{Li}u_{Lj}\rangle}{\partial x_k}\right) \tag{4.20}$$

$$\langle \theta_L u_{Li}u_{Lk}\rangle = -\frac{1}{2\tau + \tau_\Theta}\cdot\left(\tau_\Theta D_{Lik}\frac{\partial\langle u_{Li}u_{Lj}\rangle}{\partial x_k} + 2\tau_\Theta D_{Lk}\frac{\partial\langle\theta_L u_{Li}\rangle}{\partial x_k}\right) \tag{4.21}$$

Algebraic relations in Eqs. (4.19)–(4.21) were obtained at ignored small components that determined the convective transport and production of the 3rd moment of velocity fluctuations from the dispersed phase mean velocity.

Model for heat and mass transfer calculation of single evaporating droplet and conductive heat transfer between the heated wall and deposited droplets from the two-phase jet described in detail in the Sect. 2.

## 4.5 Boundary Conditions

At the outer edge of the boundary layer, the following conditions were set:

$$U = U_1; V = V_1; T = T_1; K_V = K_{V1}; k = k_1; \tilde{\varepsilon} = \tilde{\varepsilon}_1; U_L = U_{L1}; V_L = V_{L1};$$
$$M_L = M_{L1}; T_L = T_{L1};$$
$$d = d_1; \langle u_{Li}u_{Lj}\rangle =_{Li} u_{Lj}\rangle_1; \langle\theta_L u_{Lj}\rangle = \langle\theta_L u_{Lj}\rangle_1; \langle\theta_L^2\rangle = \langle\theta_L^2\rangle_1$$

$$\tag{4.22}$$

At the wall, no-slip and impermeability conditions were adopted for the gas-phase velocity:

$$U = V = \frac{\partial K_V}{\partial r} = k = 0; \; T = T_W . \tag{4.23}$$

The boundary conditions for the mean axial and radial velocities of the dispersed phase, and the boundary conditions for the pulsating components of those velocities are (Derevich 2002; Zaichik 1999)

$$\left[ \frac{1 - \chi k_t}{1 + \chi k_t} \left( \frac{2}{\pi} \langle v_L^2 \rangle \right)^{1/2} - V_{LW} \right] U_L = \langle u_{Li} u_{Lj} \rangle_W, \, V_{LW} = \frac{1 - \chi}{1 + \chi} \left( \frac{2}{\pi} \langle v_L^2 \rangle \right)^{1/2},$$

$$\left[ \frac{1 - \chi k_t^2}{1 + \chi k_t^2} \left( \frac{2}{\pi} \langle v_L^2 \rangle \right)^{1/2} - V_{LW} \right] \langle u_L^2 \rangle = -\frac{\tau}{3} \langle v_L^2 \rangle \frac{\partial \langle u_L^2 \rangle}{\partial y},$$

$$\left[ \frac{1 - \chi k_n^2}{1 + \chi k_n^2} 2 \left( \frac{2}{\pi} \langle v_L^2 \rangle \right)^{1/2} - V_{LW} \right] \langle v_L^2 \rangle = -\tau \langle v_L^2 \rangle \frac{\partial \langle v_L^2 \rangle}{\partial y},$$

$$\left[ \frac{1 - \chi k_\theta}{1 + \chi k_\theta} \left( \frac{2}{\pi} \langle v_L^2 \rangle \right)^{1/2} - V_{LW} \right] |T_W - T_L| = \langle \theta_L u_{Lj} \rangle$$

$$\left[ \frac{1 - \chi k_\theta^2}{1 + \chi k_\theta^2} \left( \frac{2}{\pi} \langle v_L^2 \rangle \right)^{1/2} - V_{LW} \right] \langle \theta_L^2 \rangle = - \left( \frac{1}{\tau} - \frac{2}{\tau_\Theta} \right)^{-1} \langle v_L^2 \rangle \frac{\partial \langle \theta_L^2 \rangle}{\partial y}.$$

$$\tag{4.24}$$

Here, $\chi$ is a coefficient that defines the probability of particle absorption by the wall (with $\chi = 1$, as particles impinge onto the wall, they all return into the stream (impermeable surface), while with $\chi = 0$ the particles, as they reach the wall, disappear from the flow ("perfectly absorbing surface")); $k_t$ and $k_n$ are coefficients that define the degree of particle momentum restoration respectively in streamwise and cross-flow direction; $k_\theta$ is a coefficient that defines the degree to which the particle temperature restores after the particle contacts the wall (with $k_\theta = 1$, the heat transfer between the particle and the wall is zero, while with $k_\theta = 0$, the reflected particles acquire a temperature equal to wall temperature); $\tau_\Theta = C_{pL} \rho_L d^2 / (12 \lambda Y)$ is the thermal relaxation time of liquid droplets; $Y = (1 + 0.3 \mathrm{Re}_L^{1/2} \, \mathrm{Pr}^{1/3})$; $g_{ut} = \frac{\tau \left( 1 - \exp(-\Omega^{tL}/\tau) \right) + \tau_\Theta (1 - f_{ut})}{(\tau + \tau_\Theta)}^{-1}$ is a function that defines the degree of particle entrainment into gas eddies and temperature pulsations (Derevich 2002); and the subscript $W$ denotes parameter values at the wall.

The system of governing Eqs. (3.20)–(4.3) with boundary conditions (4.4)–(4.6) describes processes in a gas-droplet flow; this system allows one to calculate all quantities of interest, including the distributions of temperature and concentrations of the vapor–gas mixture phases and components, to trace the evolution of particle sizes, and to evaluate the degree of heat-transfer intensification due to evaporation processes.

## 4.6 Numerical Procedure and Model Testing

The numerical solutions based on a finite-volume method (Patankar 1980) of the parabolic equations mean momentum, energy, vapor diffusion into the binary air–steam mixture and turbulent transport equations for both phases in the boundary layer approach. We used non-uniform staggered grid in longitudinal and transverse directions. It was used a very fine grid close to the wall. The QUICK scheme (Leonard 1979) was used for convective fluxes and central differences were utilized for diffusion fluxes. The SIMPLEC solution algorithm (Van Doormaal and Raithby 1984) was adopted for pressure–velocity coupling.

The basic grid being used had $300 \times 100$ control volumes (CVs) with a higher resolution in the wall and initial regions. The first cell is located at distance $y_+ = yU_*/\nu = 0.3$–$0.5$ from the wall, where $U_*$ is the friction velocity. At least 10 CVs has been generated to be able to resolve the mean velocity field and turbulent quantities in the viscosity-affected near-wall region ($y_+ < 10$). The mesh configuration was tested to be sufficient to provide grid independent results. Additionally, a series of test computations for a gas-droplet flow occupying a total of $500 \times 200$ CVs was performed. Grid sensitivity studies are carried out to determine the optimum grid resolution that gives the mesh independent solution. For all further investigations, a grid with $300 \times 100$ control volumes in longitudinal and transversal directions in these predictions has been used. The grid convergence is checked by three cases: $100 \times 50$ control volumes, $200 \times 75$, and $400 \times 200$. The results were a change less than 1 % in local Nusselt number.

Differences in the value of heat transfer calculated for a gas-droplets jet flow less than 1 %. The solutions presented here are considered grid independent. The maximum error $e_{\max}$ has been defined as

$$e_{\max} = \max_{i=1,N} \left| \phi_i^n - \phi_i^{n-1} \right| \leq 10^{-6}, \tag{4.25}$$

where $N$ denotes the total number of control volumes, subscript $i$ denotes the specific CV, superscript $n$ denotes the iteration level and $\phi$ denotes all flow variables.

In our computations we used several $k-\varepsilon$ turbulence models (Hwang and Lin 1998; Jones and Launder 1973; Myong and Kasagi 1990). The results are compared with the experimental study of regularities of wall friction and heat transfer in single-phase turbulent boundary layer (Kutateladze and Leont'ev 1989; Schlichting 1960). The better agreement is obtained with the use of model (Hwang and Lin 1998).

## 4.7 Numerical Results and Discussions

All numerical calculations were performed for a mist gas-droplets flow. The gas velocity in free stream was assumed to be spatially uniform, equal to $U_1 = 10$ m/s, and the temperature of the mixture was also constant, $T_1 = 293$ K. For the mass vapor content of outside flow, a value $M_{V1} = 0.005$ was adopted. In the calculations, four quantities were varied parameters: the Reynolds number $Re_x = U_1 x / v = 3 \times 10^5 - 10^6$, the wall temperature $T_W = 323 - 473$ K, the mass concentration of the liquid phase $M_{L1} = 0 - 0.05$, and the free-stream droplet diameter $d_1 = 0 - 100$ μm. All calculations were performed for identically sized water droplets. No droplets evaporation was assumed to occur in the free flow.

Main attention in subsequent calculations was focused on the effect of droplet size on the distributions of gas velocities and temperature, and also on the distributions of the concentrations of the two-phase mixture components in the boundary layer. The calculated normalized velocity and temperature profiles across the boundary layer are shown in Fig. 4.1; here, the dotted curve shows data calculated for a single-phase air flow, $\Theta = (T_W - T)/(T_W - T_1)$ is the temperature profile in the gas-droplets flow, and $\delta^* = \int_0^\delta \left(1 - \frac{\rho U}{\rho_1 U_1}\right) dy$ (Kutateladze and Leont'ev 1989; Schlichting 1960) is the displacement thickness. In the absence of liquid phase ($M_{L1} = 0$) the calculated data for the gas-flow velocity (Fig. 4.1a) were found to follow the power-law profile $U/U_1 = (y/\delta)^{1/7}$ (Kutateladze and Leont'ev 1989; Schlichting 1960). An increase in liquid concentration results in that the velocity profile in the boundary layer becomes more filled due to evaporation processes proceeding in the vicinity of the wall. It should be noted, however, that the presence of the liquid phase in the flow was found to affect the temperature distributions to a much greater extent (Fig. 4.1b), finally causing a more intense growth of heat-transfer intensity in comparison with friction. The data in Fig. 4.1a show that, on variation of droplet diameter, the boundary-layer thickness remains almost unchanged. Note that particles initially sized 50 μm exert a more pronounced influence on the gas-phase temperature, whereas the action on gas velocity and, hence, on wall friction increases almost linearly with particle diameter.

The mean gas velocity profiles obtained at $Re_x = 5 \times 10^5$ for the one-phase and (curve *1*) droplets laden flow (curves 2 and 3) are presented in Fig. 4.2, which is a logarithmic presentation of the dimensionless velocity $U_+ = U/U_*$ as a function of the dimensionless wall distance $y_+ = yU_*/v$. Bold lines are the log law of the wall (Schlichting 1960)

$$\begin{cases} U_+ = y_+ \\ U_+ = 5.75 \log(y_+) + 5.5. \end{cases} \tag{4.26}$$

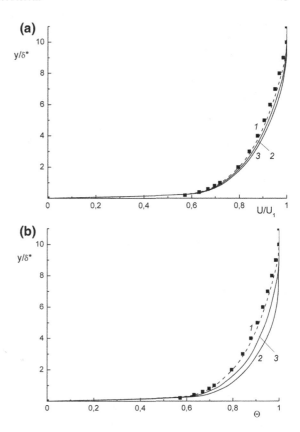

**Fig. 4.1** Velocity (**a**) and temperature (**b**) profiles of the gas phase on the transverse coordinate. $Re_x = 5 \times 10^5$, $\delta^* = 2$ mm, $U_1 = 10$ m/s, $T_W = 373$ K, $M_{L1} = 5$ %. Points are the results of (Pakhomov et al. 2007). $1$— $d_1 = 0$ μm; $2$—50; $3$—100

Comparison between computed and well-known classical log law of the wall velocity profiles shows that the flow is not significantly altered by the variation of droplets mass fraction (see Fig. 4.2a) and its initial size (see Fig. 4.2b).

The above specific features revealed in the velocity and temperature profiles of the two-phase flow past a vertical plate were also reflected in the distribution of droplet sizes in the boundary layer. These data are shown in Fig. 4.3. With increasing in the initial size, droplets undergo less intense evaporation in the wall zone. As the stream moves farther downstream, the zone occupied by the single-phase vapor–gas flow in the vicinity of the wall grows in thickness.

The distributions of the mass concentration of the components across the boundary layer differ in shape from the velocity and temperature distributions. Figure 4.4 shows the profiles of the mass fraction of liquid (Fig. 4.4a) and vapor (Fig. 4.4b). Due to boundary-layer heating and evaporation processes, the mass concentration of the liquid phase permanently decreases as we approach the wall (Fig. 4.4a). At certain droplet-size values in the flow core and at a certain wall temperature, the wall zone becomes free of droplets, and a zone occupied by a single-phase vapor-air flow forms there. As the stream moves farther downstream, the near-wall zone of the single-phase flow in the vicinity of the wall grows

**Fig. 4.2** Effect of the
droplets initial size on the
mean gas velocity profiles.
*Bold lines* represent the law
of the wall by Eq. (4.26).
$Re_x = 5 \times 10^5$,
**a** $d_1 = 50$ μm. *1* $M_{L1} = 0$; *2*
5 %; *3* 10 %. **b** $M_{L1} = 5$ %.
*1*—$d_1 = 0$ μm; *2*—50; *3*—
100

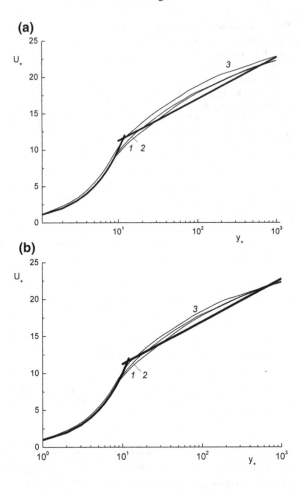

**Fig. 4.3** Distributions of
droplets diameter across the
boundary layer. *Solid lines*
$Re_x = 5 \times 10^5$, $\delta^* = 2$ mm,
*dashed lines* $Re_x = 10^6$,
$\delta^* = 3$ mm. $U_1 = 10$ m/s,
$T_W = 373$ K, $M_{L1} = 5$ %.
*1*—$d_1 = 10$ μm; *2*—50; *3*—
100

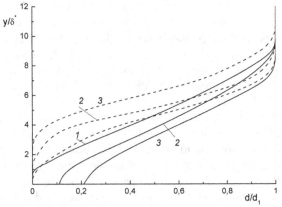

**Fig. 4.4** Profiles of mass fraction of droplets (**a**) and vapor (**b**) across the boundary layer. Key and conditions of numerical modeling, see Fig. 4.3

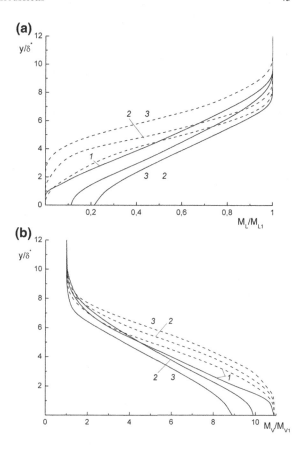

thicker; this is evident from Fig. 4.3. With decrease of the mass fraction of droplets in the free stream $M_{L1}$, and also with increase of wall temperature, the single-phase flow zone starts forming earlier, exhibiting also a more pronounced growth in terms of its downstream extension and width.

As the stream moves along the plate surface, the vapor concentration increases due to evaporation of the dispersed phase (see Fig. 4.4b). The increase in vapor concentration may appear rather considerable in comparison with the vapor concentration in the approaching flow ($M_V/M_{V1} > 10$). A decrease of droplet size causes a reduction of vapor concentration in the boundary layer due to a profound reduction of the inter-phase surface area. In the outside of boundary layer, the mass concentrations of droplets and water vapor are roughly the same as the concentrations in the free-stream flow, which result can be attributed to insignificant heating of this flow zone and, hence, to a low intensity of evaporation processes.

The variation of mean gas flow turbulence along the transverse coordinate is shown in Fig. 4.5a, where $k_A$ is the turbulent kinetic energy (TKE) in the single-phase flow. It should be noted that addition of readily evaporating droplets reduces the turbulence level of the gas flow, this effect becoming more pronounced with

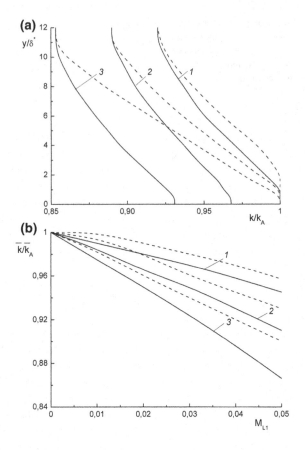

increasing the dispersed-phase size. Thus, small particles readily get entrained into
the turbulent gas motion and take from it some part of turbulent energy. In the wall
zone, even at a small distance from the inlet to the computational domain
($Re_x = 5 \times 10^5$), we have $k/k_A \approx 1$, because this flow region contains no liquid
droplets, which has already evaporated completely in the upstream zone, making
the turbulence level here approach the value typical of single-phase flow. In the
external flow, which still contains liquid droplets, we have: $k/k_A \rightarrow$ const. As the
flow moves downstream, the degree of TKE modification decreases as particles
undergo evaporation and gas turbulence grows in value, so that $k/k_A \rightarrow 1$, this
effect being manifested at a greater distance from the plate surface.

Figure 4.5b shows the values of the gas-phase TKE averaged over the cross-
section of boundary layer at two different distances from the inlet to the calculation

domain. The mean turbulence was calculated as $\bar{k} = \frac{1}{\delta} \int\limits_0^{\delta} k\,dy$. Here, $\bar{k}_A$ is the mean

turbulence kinetic energy in the single-phase flow. Thus, we showed that the
degree of TKE modification is defined not only by the initial droplet diameter $d_1$,

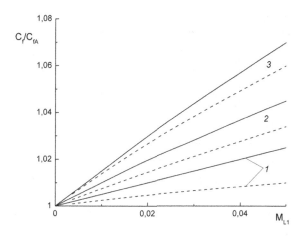

**Fig. 4.6** Profiles of mean friction coefficient in mist flow. *Solid lines* are for $Re_x = 5 \times 10^5$, *dashed lines* are for $Re_x = 10^6$. $U_1 = 10$ m/s, $T_W = 373$ K. *1—$d_1 = 10$ μm; 2—50; 3—100*

but also by the droplets mass concentration $M_{L1}$ in the free stream. A characteristic feature here is a profound suppression of gas-flow turbulence kinetic energy at small distances from the inlet to the calculation domain, consistent with the data of Fig. 4.5a.

The specific features observed in the evolution of local characteristics of the gas-droplets flow (see Figs. 4.1, 4.2, 4.3, 4.4, 4.5) affect the downstream evolution of friction and heat transfer. As it is seen from the data of Fig. 4.6, the liquid phase present in the flow rather weakly affects the evolution of the friction coefficient (which varies here within 8 %). Here, $C_{fA}$ is the wall friction coefficient in the single-phase flow. This result is consistent with the data of Fig. 4.1b, where a change of particle size also exerted an insignificant influence on the variation of velocity profile. Note that the predicted dependences for friction coefficient obtained for $M_{L1} = 0$ well agree with the following dependence reported in (Kutateladze and Leont'ev 1989; Schlichting 1960):

$$C_f/2 = 0.022 Re_x^{-0.2}. \tag{4.27}$$

For heat transfer, quite a different picture is observed (see Fig. 4.7). The presence of liquid droplets in the flow leads to a considerable increase of the heat transfer intensification ratio. Here, $Nu_A$ is the heat transfer coefficient in the single-phase flow. This ratio can reach rather high values (approximately four in comparison with the single-phase flow). With increasing the droplet size the heat transfer intensity increases, this effect being observed more clearly at small distances from the inlet section.

In the Fig. 4.8 is presented the effect of initial droplets diameter on the heat transfer enhancement ratio $Nu/Nu_A$ along the flat plate length. It is seen that the effect of variation of droplets size on heat transfer has complex behavior than the change of droplets mass fraction. Initially the increase of droplets diameter ($d_1 \leq 50$ μm) causes the augmentation of heat transfer rate due to its evaporation near the wall surface. But for the relatively large droplets ($d_1 > 70$ μm) it is

**Fig. 4.7** The influence of droplets mass fraction on heat transfer enhancement ratio between plate and gas-droplets flow. Key and conditions of numerical modeling, see Fig. 4.6

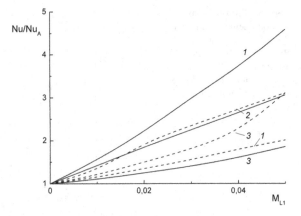

**Fig. 4.8** The effect of initial droplets diameter on heat transfer enhancement ratio. $1$—$\mathrm{Re}_x = 5 \times 10^5$; $2$—$8 \times 10^5$; $3$—$10^6$, $4$—$2 \times 10^6$

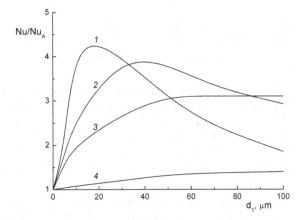

observed the opposite trend. The value of heat transfer is decreased due to significant reduction of the droplets surface. The large droplets have the smaller heat transfer enhancement ratio, but they evaporate on the large distance from the beginning of the plate (see line $4$). In particular this leads to that the largest droplets have the higher value of heat transfer rate for the distance $\mathrm{Re}_x = 2 \times 10^6$.

The effects due to the initial concentration of liquid droplets, their diameter, and wall temperature on the distribution of Nusselt number over the plate length are illustrated by Fig. 4.9, respectively a, b, and c. At all other conditions kept unchanged, an increase in the Reynolds number $\mathrm{Re}_x$ of the two-phase flow leads to pronounced heat-transfer intensification. A similar tendency was also traced for other regime conditions. With increasing the mass concentration of the liquid phase the correlation curves $\mathrm{Nu}_x = f(\mathrm{Re}_x)$ ascend more rapidly (see Fig. 4.9a), and in the downstream region, the height of the wall zone occupied by the single-phase flow was thicker, the slope of those dependences approaches the law of turbulent

**Fig. 4.9** The distributions of Nusselt number along the plate coordinate for various initial droplets mass fraction (**a**), droplets initial diameter (**b**) and wall temperature (**c**). Points are calculation by the formula (4.28). *Dashed lines* are the prediction by authors for one-phase air flow. $U_1 = 10$ m/s, $d_1 = 50$ μm.
**a** $T_W = 373$ K, $1—M_{L1} = 0$; $2—0.01$; $3—0.02$; $4—0.05$.
**b** $M_{L1} = 0.05$, $1—d_1 = 0$ μm; $2—10$; $3—50$; $4—100$.
**c** $M_{L1} = 0.05$, $1—T_W = 323$ K; $2—373$; $3—423$

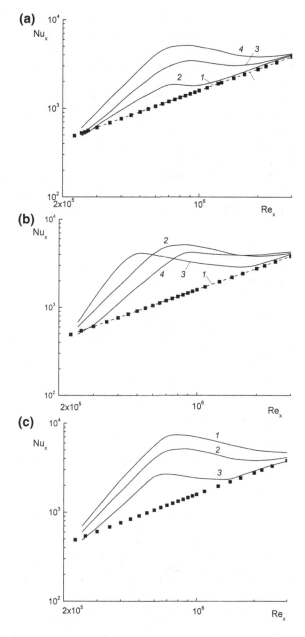

boundary layer under standard conditions (Kutateladze and Leont'ev 1989; Schlichting 1987):

$$\mathrm{Nu}_x = 0.0288 \mathrm{Re}_x^{0.8} \mathrm{Pr}^{0.4}. \tag{4.28}$$

An important parameter that defines the intensity of heat-transfer processes in two-phase flows is the size of liquid droplets in the free stream. Results gained in such calculations are demonstrated in Fig. 4.9b. The evaporation process proceeds at a higher rate in the case of small particle sizes ($d_1 = 10$ μm) yet at a small distance from the inlet section. Farther downstream, as droplets undergo evaporation, the curves $Nu_x = f(Re_x)$ start closely following the curves for single-phase flow. With increasing the diameter of liquid droplets while keeping the mass content of the liquid unchanged, the area of the inter-phase surface decreases considerably. As a result, the rate of heat transfer between the plate and the gas-droplet flow falls in value.

The wall temperature has a profound influence on the degree of heat-transfer intensification (see Fig. 4.9c). As the wall temperature increases, the near-wall zone gets more rapidly freed from the liquid phase, and the rate of heat transfer shows a decrease.

# References

X.Q. Chen, Heavy particles dispersion in inhomogeneous, anisotropic turbulent flows. Int. J. Multiphase Flow **26**, 635–661 (2000)

R. Clift, J.R. Grace, M.E. Weber, *Bubbles, Drops and Particles*. (Academic Press, New York, 1978)

C.T. Crowe, M. Sommerfeld, T. Tsuji, *Fundamentals of Gas-Particle and Gas-Droplet Flows* (CRC Press, Boca Raton, 1998)

I.V. Derevich, The hydrodynamics and heat transfer and mass transfer of particles under conditions of turbulent flow of gas suspension in a pipe and in an axisymmetric jet. High Temp. **40**, 78–91 (2002)

I.V. Derevich, L.I. Zaichik, Particle deposition from a turbulent flow. Fluid Dyn. **23**, 722–729 (1988)

D.A. Drew, Mathematical modeling of two-phase flow. Ann. Rev. Fluid Mech. **15**, 261–291 (1983)

D.P. Healy, J.B. Young, Full Lagrangian methods for calculating particle concentration fields in dilute gas-particle flows. Proc. Royal Society A. **461**, 2197–2225 (2005)

C.B. Hwang, C.A. Lin, Improved low-Reynolds-number $k-\tilde{\varepsilon}$ model based on direct simulation data. AIAA J. **36**, 38–43 (1998)

M. Ishii, *Thermo-Fluid Theory of Two-Phase Flows* (Eyrolles, Paris, 1975)

W.P. Jones, B.E. Launder, The calculation of low-Reynolds-number phenomena with a two-equation model of turbulence. Int. J. Heat Mass Transfer **15**, 1119–1130 (1973)

S.S. Kutateladze, A.I. Leont'ev, *Heat and Mass Transfer in Turbulent Boundary Layer*. (Hemisphere, New York, 1989)

B.P. Leonard, A stable and accurate convective modelling procedure based on quadratic upstream interpolation. Comp. Methods Appl. Mech. Eng. **19**, 59–98 (1979)

C. Marchioli, M. Picciotto, A. Soldati, Influence of gravity and lift on particle velocity statistics and transfer rates in turbulent vertical channel flow. Intl J. Multiphase Flow **33**, 25–227 (2007)

H.K. Myong, N. Kasagi, A new approach to the improvement of $k$-$\varepsilon$ turbulence model for wall-bounded shear flows. Int. J. JSME. Ser. II **33**, 63–72 (1990)

A.N. Osiptsov, Lagrangian modeling of dust admixture in gas flows. Astrophys. Space Sci. **274**, 377–386 (2000)

M.A. Pakhomov, V.I. Terekhov, Enhancement of an impingement heat transfer between turbulent mist jet and flat surface. Int. J. Heat Mass Transfer **53**, 3156–3165 (2010)

M.A. Pakhomov, V.I. Terekhov, Modeling of the flow structure and heat transfer in a gas-droplet turbulent boundary layer. Fluid Dyn. **47**, 168–177 (2012)

M.A. Pakhomov, M.V. Protasov, V.I. Terekhov, AYu. Varaksin, Experimental and numerical investigation of downward gas-dispersed turbulent pipe flow. Int. J. Heat Mass Transfer **50**, 2107–2116 (2007)

S.V. Patankar, *Numerical Heat Transfer and Fluid Flow*. (Hemisphere, Washington, 1980)

C.B. Rogers, J.K. Eaton, The behavior of small particles in a vertical turbulent boundary layer in air. Int. J. Multiphase Flow **16**, 819–834 (1990)

H. Schlichting, *Boundary Layer Theory*. (McGraw-Hill Publishing House, New York, 1960)

V.I. Terekhov, M.A. Pakhomov, Numerical study of heat transfer in a laminar mist flow over an isothermal flat plate. Int. J. Heat Mass Transfer **45**, 2077–2085 (2002)

V.I. Terekhov, M.A. Pakhomov, Numerical simulation of hydrodynamics and convective heat transfer in turbulent tube mist flow. Int. J. Heat Mass Transfer **46**, 1503–1517 (2003)

V.I. Terekhov, M.A. Pakhomov, Film-cooling enhancement of the mist vertical wall jet on the cylindrical channel surface with heat transfer. Trans. ASME J. Heat Transfer **131**, Paper 062201 (2009a)

V.I. Terekhov, M.A. Pakhomov, Predictions of turbulent flow and heat transfer in gas-droplets flow downstream of a sudden pipe expansion. Int. J. Heat Mass Transfer **52**, 4711–4721 (2009b)

V.I. Terekhov, M.A. Pakhomov, A.V. Chichindaev, Effect of evaporation of liquid droplets on the distribution of parameters in a two-species laminar flow. J. Appl. Mech. Techn. Phys. **41**, 1020–1028 (2000)

V.I. Terekhov, M.A. Pakhomov, K.A. Sharov, N.E. Shishkin, The thermal efficiency of near-wall gas-droplets screens. II. Experimental study and comparison with numerical results. Int. J. Heat Mass Transfer **48**, 1760–1771 (2005)

J.P. Van Doormaal, G.D. Raithby, Enhancements of the SIMPLE method for predicting incompressible fluid flow. Int. J. Numerical Heat Transfer A **7**, 147–164 (1984)

A.Yu. Varaksin, *Turbulent Particles-Laden Gas Flows* (Springer, Berlin, 2007)

J. Wang, E.K. Levy, Particle behavior in the turbulent boundary layer of a dilute gas-particle flow past a flat plate. Int. J. Exp. Fluid Sci. **30**, 473–483 (2006)

L.I. Zaichik, A statistical model of particle transport and heat transfer in turbulent shear flows. Phys. Fluids A **11**, 1521–1534 (1999)

L.I. Zaichik, V.M. Alipchenkov, E.G. Sinaiski, *Particles in Turbulent Flows* (Wiley-VCH, Berlin, 2008)

# Chapter 5
# Comparison with Experimental Data in a Flat Plate Turbulent Gas-Particles Boundary Layer

Unlike for laminar boundary layer, we failed to come across in the literature measured data concerning the flow structure and heat transfer in turbulent mist gas-droplets flows over the flat vertical plate. That is why in the present section is given a comparison of our calculated data with available well-known experimental data on the structure of turbulent boundary layer air flows laden with solid particles (Rogers and Eaton 1990, 1991).

The experimental studies reported in (Rogers and Eaton 1990, 1991) are carried out using an LDA instrument adapted for measurement of two-phase flows. In the paper (Rogers and Eaton 1990), measurements are carried out for identically sized glass particles with $d = 50$ and 90 μm at a mass concentration of particles in the ascending flow $M_{L1} = 0.02$. In (Rogers and Eaton 1991) experimentally examines the effect due to copper particles of diameter $d = 70$ μm ($M_{L1} = 0.2$) on gas-phase turbulence modification. The experiments are performed for distances $x = 55$ cm ($\delta = 20$ mm, $\delta^* = 3$ mm, $\delta^{**} = 2.1$ mm, $\mathrm{Re}_x = 2.9 \times 10^5$) and 85 cm ($\delta = 24$ mm, $\delta^* = 3.8$ mm, $\delta^{**} = 2.6$ mm, $\mathrm{Re}_x = 4.5 \times 10^5$) from the leading edge of the plate streamlined by a downward two-phase air flow seeded with solid particles. The gas velocity in the free flow is $U_1 = 8$–8.2 m/s. Note that, in the experiments, the gas velocities for the single-phase and two-phase mixtures are identical.

The profiles of mean longitudinal gas velocity at the distances 55 cm (a) and 85 cm (b) from the leading edge of the plate are shown in Fig. 5.1. Note an insignificant difference in the profile of longitudinal gas velocity and in the boundary layer thickness at both distances in comparison with single-phase flow; this observation provides agreed with to our numerical results is shown in Fig. 4.1.

Figure 5.2 shows the profile of mean longitudinal gas velocity in the single-phase flow (curve 1) and the profiles of particle velocity for differently sized particles (curves 2 and 3) over the boundary layer cross-section. In contrast to the data of Fig. 5.1b, here the gas velocity values are greater than the particle velocities throughout the whole boundary layer cross-section. Since in the calculations we treated an ascending flow, it is apparent that the velocity of 90-μm diameter dispersed phase should be lower than that of 50-μm particles.

V. I. Terekhov and M. A. Pakhomov, *Flow and Heat and Mass Transfer in Laminar and Turbulent Mist Gas-Droplets Stream over a Flat Plate*, SpringerBriefs on Multiphase Flow, DOI: 10.1007/978-3-319-04453-8_5,

**Fig. 5.1** Streamwise gas phase mean velocity profiles at $x = 55$ cm (**a**) and $x = 85$ cm (**b**). Points are the measurements of (Rogers and Eaton 1991), *lines* are the predictions by this chapter. *1*—single-phase boundary layer; *2*—two-phase boundary layer laden with copper particles $d = 70$ μm. $M_{L1} = 0.2$

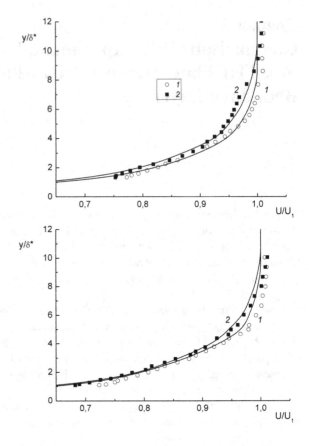

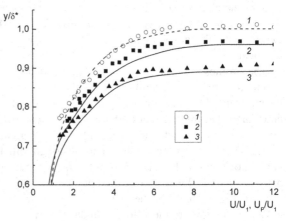

**Fig. 5.2** Streamwise mean gas velocity (*dashed curves*) and particles (*solid curves*) on the two-phase gas-dispersed boundary layer. Points are the measurements by (Rogers and Eaton 1990), lines are the predictions by this chapter. $M_{L1} = 0.02$, $x = 55$ cm. *1*—gas phase; *2*—glass particles ($d = 50$ μm); *3*—glass particles ($d = 90$ μm)

From the data in Figs. 5.1 and 5.2 it is seen that, here, we observe rather a good coincidence between the values of longitudinal velocities of the phases measured in (Rogers and Eaton 1990, 1991) in the ascending and descending flows over a

**Fig. 5.3** Distributions of streamwise (**a**) and transverse (**b**) velocity fluctuations of gas (*solid lines*) and dispersed phases (*dashed lines*) in the boundary layer. Key and conditions of numerical modeling, see Fig. 5.2. $M_{L1} = 0.02$, $x = 55$ cm

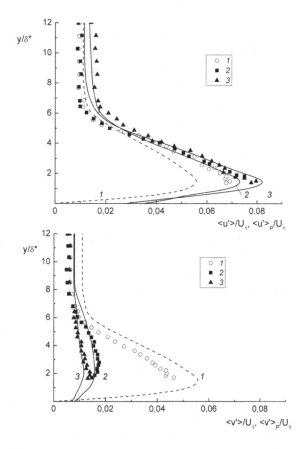

wide range of particle materials and sizes (respectively the Stokes number St and the Reynolds number $\mathrm{Re}_L$ of dispersed phase).

The variation of the longitudinal (a) and transverse (b) turbulence components of the gas phase (in single-phase flow) and that of the dispersed phase over the transverse coordinate are shown in Fig. 5.3. In the longitudinal direction the intensity of particle velocity pulsations $\langle u_p^{1/2} \rangle / U_1$ (curves 2 and 3) is greater that the same value for the gas, $\langle u^{1/2} \rangle / U_1 = 2k/(3U_1)$ (curve 1), throughout the whole boundary-layer cross-section. This result is consistent with the experimental and numerical data obtained in (Derevich 2002; Pakhomov and Terekhov 2010, 2012; Varaksin 2007; Varaksin et al. 1995, 2011) for turbulent gas-dispersed pipe flows. All of the turbulent time scales decrease on approaching the wall. It is the high gradients of mean axial gas velocity which are mainly responsible for abrupt increase in the intensity of axial fluctuations of particle velocity in the near wall region. The nonuniformity of the axial velocity profile of the carrier phase causes the nonuniformity of the profile of averaged axial velocity of particles.

**Fig. 5.4** Streamwise fluctuations of the gas velocity. Points are the measurements by (Rogers and Eaton 1991), *lines* are the predictions by this chapter. $M_{L1} = 0.2$, $x = 85$ cm, $Re_x = 4.35 \times 10^5$. *1*— single-phase flow; *2*—two-phase flow with 70 μm copper particles

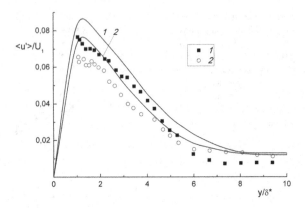

The distributions of the radial velocity fluctuations intensity of one-phase air flow $\langle v^{1/2} \rangle / U_1 = 2k/(3U_1)$ and particles $\langle v_P^{1/2} \rangle / U_1$ given in the Fig. 5.3b. The intensity of radial velocity fluctuations of particle is lower than one for the air of unladen flow. These facts may be interpreted as follows. The Stokes number in a large-scale fluctuating motion for the considered conditions is St > 1, and it follows that particles are involved into the large-scale fluctuation motion and extract the energy of turbulent eddies of the carrier phase. A decrease in the intensity of transverse fluctuations of the gas phase leads to a suppression in the fluctuations of particles. In the near wall region, the intensity of fluctuations of gas and particles increases abruptly. The diameter of particles has a strong effect on the intensity of fluctuations in the near wall region of the boundary layer. The turbulence damping being greater for the larger particles with longer relaxation time. This is demonstrated by experimental and numerical results.

The comparison of measured (Rogers and Eaton 1991) and computed by authors streamwise turbulence intensities in the single-phase and two-phase flows across the boundary layer at $x = 85$ cm were presented in the Fig. 5.4. It is seen substantial decrease of turbulence intensity in the two-phase flow especially in the near wall zone.

## References

C.B. Rogers, J.K. Eaton, The behavior of small particles in a vertical turbulent boundary layer in air. Int J Multiph Flow **16**, 819–834 (1990)

C.B. Rogers, J.K. Eaton, The effect of small particles on fluid turbulence in a flat plate, turbulent boundary layer in air. Phys Fluids A **3**, 928–937 (1991)

I.V. Derevich, The hydrodynamics and heat transfer and mass transfer of particles under conditions of turbulent flow of gas suspension in a pipe and in an axisymmetric jet. High Temp **40**, 78–91 (2002)

M.A. Pakhomov, V.I. Terekhov, Enhancement of an impingement heat transfer between turbulent mist jet and flat surface. Int J Heat Mass Transf **53**, 3156–3165 (2010)

M.A. Pakhomov, V.I. Terekhov, Modeling of the flow structure and heat transfer in a gas-droplet turbulent boundary layer. Fluid Dyn **47**, 168–177 (2012)

A.Yu. Varaksin, *Turbulent Particles-Laden Gas Flows* (Springer, Berlin, 2007)

A.Yu. Varaksin, D.S. Mikhatulin, YuV Polezhaev, A.F. Polyakov, Measurements of velocity fields of gas and solid particles in the boundary layer of turbulized heterogeneous flow. High Temp **33**, 911–917 (1995)

A.Yu. Varaksin, M.E. Romash, V.N. Kopeitsev, M.A. Gorbachev, Physical simulation of air tornados: some dimensionless parameters. High Temp **49**, 310–313 (2011)

# Conclusions

The effect of droplets evaporation in laminar and turbulent boundary layers of a dilute two-phase flow on a vertical flat plate is numerically studied. The numerical simulation is based on Eulerian two-fluid model. The low-Reynolds number $k-\varepsilon$ model is used to account for gas phase turbulence in the boundary layer.

As a first approximation, in the case of small dispersed phase sizes, in studying the flow structure and heat-and mass transfer processes in laminar boundary-layer flows at small particle sizes one can use a single-velocity model.

A significant anisotropy of fluctuations of particle velocity has been found. The amplitude of turbulent fluctuations of particle velocity in the axial direction is higher than that in the radial direction. In addition to being associated with the inherent anisotropy of turbulence of the gas phase, this effect is caused by additional generation of turbulence during the motion of particles in the field of gradient of averaged axial velocity of the dispersed phase. The intensity of fluctuations of particle velocity in the axial direction may be higher than one in the case of the gas phase. The axial and radial fluctuations of particle velocity depend strongly on the particle mass concentration and their diameter. Loadings of solid particles into the gas flow causes a decrease in the level of turbulence of the gas phase because of the involvement of particles into fluctuation motion.

The thickness of the boundary layer is almost independent of the inlet droplet size and mass concentration. The presence of the liquid phase much greater affects the temperature distribution, which ultimately causes a more rapid growth of heat fluxes, as compared to the friction. With increase in the particle concentration, the friction on the wall increases only slightly.

The effects of evaporating droplets on friction and heat transfer in two-phase flow are different. The friction increases insignificantly with increasing initial size and mass concentration of the dispersed phase, whereas heat transfer shows a non-monotonic behavior, first showing an increase with particle size and, then, exhibiting a decrease as the inter-phase surface area markedly decreases. This conclusion is also made in our simulation studies of heat transfer in turbulent gas-droplets pipe flows, wall mist jet and impinging gas-droplets jet. The intensity of heat transfer in a turbulent gas-droplets flow depends significantly on the free-stream velocity, the droplets diameter, and the wall temperature. It is shown that the increase in the liquid phase mass concentration results in a considerable

V. I. Terekhov and M. A. Pakhomov, *Flow and Heat and Mass Transfer in Laminar and Turbulent Mist Gas-Droplets Stream over a Flat Plate,* SpringerBriefs on Multiphase Flow, DOI: 10.1007/978-3-319-04453-8,

intensification of heat transfer, as compared with single-phase flow. The variation of the flow velocity or the longitudinal coordinate at fixed $Re_x$ results in different dependences of heat transfer, which indicates a complex and ambiguous influence of $Re_x$ on heat transfer in gas-droplets flow both in laminar and turbulent boundary layers.

The numerical results are compared with well-known experimental studies of Heyt and Larsen (1970), Hishida et al. (1980), Rogers and Eaton (1990, 1991) and the agreement is quite satisfactory.

# References

J.W. Heyt, P.S. Larsen, Heat transfer to binary mist flow. Int J Heat Mass Transf **14**, 1149–1158 (1970)

K. Hishida, M. Maeda, S. Ikai, Heat transfer from a flat plate in two-component mist flow. Trans ASME J Heat Transf **102**, 513–518 (1980)

C.B. Rogers, J.K. Eaton, The behavior of small particles in a vertical turbulent boundary layer in air. Int J Multiph Flow **16**, 819–834 (1990)

C.B. Rogers, J.K. Eaton, The effect of small particles on fluid turbulence in a flat plate, turbulent boundary layer in air. Phys Fluids A **3**, 928–937 (1991)

Printed in the United States
By Bookmasters